DFT: A Formula Handbook

N.B. Singh

DEDICATION

To Nature,

I dedicate this book to you, the source of all life. You are my inspiration, my teacher, and my friend.

Thank you for teaching me about the beauty of the world around me. Thank you for showing me the power of the natural world. Thank you for giving me a sense of peace and tranquillity.

I promise to do my part to protect you and your many wonders. I will teach my children about the importance of conservation and sustainability. I will work to make the world a better place for all living things.

Thank you for everything, Nature.

With love,

N.B Singh

Contents

Appendix   **89**

Preface

About This Handbook

Welcome to "DFT: A Formula Handbook." This handbook aims to provide a comprehensive and concise collection of formulas, equations, and practical examples related to Density Functional Theory (DFT). Whether you are a student, researcher, or practitioner in the field of computational chemistry and materials science, this handbook is designed to serve as a quick reference guide for your DFT-related endeavors.

Organization of the Handbook

The handbook is organized into several chapters, each focusing on a specific aspect of DFT. From the mathematical foundations to advanced topics and applications, every effort has been made to cover a wide range of topics. The chapters are designed to be self-contained, allowing you to navigate directly to the information you need.

Contents

The handbook covers the following key areas:

- Mathematical Basis of DFT

- Formalism and Equations in DFT

- Approximations in DFT

- Numerical Techniques

- Basis Sets and Pseudopotentials

- Computational Methods in DFT

- Applications of DFT

- Advanced Topics in DFT

- Emerging Trends and Future Perspectives

How to Use This Handbook

Each chapter begins with a brief introduction to the topic and is followed by a collection of formulas, equations, and examples. Mathematical formulations are presented in a clear and concise manner. Practical examples and numerical illustrations are included to aid in understanding and application.

Happy Reading!

I hope you find "DFT: A Formula Handbook" a valuable resource in your journey through the intricacies of Density Functional Theory.

Chapter 1

Introduction to DFT

1.1 Overview of Quantum Mechanics

Quantum mechanics is the theoretical framework that governs the behavior of particles at the atomic and subatomic levels. Developed in the early 20th century, quantum mechanics introduced a paradigm shift in our understanding of the physical world, challenging classical notions and providing a more accurate description of nature's fundamental building blocks.

1.1.1 Wave-Particle Duality

A cornerstone of quantum mechanics is the concept of wave-particle duality. Matter, at the quantum level, exhibits both particle-like and wave-like characteristics. This duality is evident in the famous double-slit experiment, where particles like electrons create interference patterns similar to waves.

1.1.2 Schrodinger Equation

Central to quantum mechanics is the Schrodinger equation, a differential equation that describes how the quantum state of a physical system changes over time. The solutions to this equation, known as wavefunctions, encode information about the probabilities of finding a particle in a particular state.

The time-independent Schrodinger equation for a non-relativistic particle is given by:

$$\hat{H}\psi = E\psi$$

Here, $\hat{H}$ is the Hamiltonian operator, ψ is the wavefunction, and E is the energy of the system.

1.1.3 Quantum States and Observables

Quantum states are described by wavefunctions, and observables (physical properties like position and momentum) are represented by operators. The act of measurement in quantum mechanics involves the collapse of the wavefunction to one of its eigenstates, corresponding to a definite value of the observable.

1.1.4 Heisenberg Uncertainty Principle

A fundamental aspect of quantum mechanics is the Heisenberg Uncertainty Principle, formulated by Werner Heisenberg. It states that certain pairs of properties, such as position and momentum, cannot be simultaneously known with arbitrary precision. The more precisely one property is measured, the less precisely the conjugate property can be determined.

1.1.5 Quantum Mechanics and DFT

In the context of density functional theory (DFT), quantum mechanics provides the theoretical foundation for understanding electronic structures. DFT is a quantum mechanical method used for calculating electronic properties of many-body systems, making it a powerful tool in materials science and computational chemistry.

1.1.6 Mathematical Formalism

The mathematical formalism of quantum mechanics involves linear algebra and functional analysis. Hilbert spaces and operators are used to represent physical states and observables, respectively. The inner product of wavefunctions in a Hilbert space provides a measure of probability amplitude.

1.1.7 Sample Equations

Let's consider a simple example of a particle in a box. The wavefunction $\psi(x)$ for a particle confined to a one-dimensional box of length L is given by:

$$\psi(x) = \sqrt{\frac{2}{L}} \sin\left(\frac{n\pi x}{L}\right)$$

where n is the quantum number.

1.1.8 Chemical Bonding

Quantum mechanics explains the nature of chemical bonds. The molecular orbital theory, a quantum mechanical model, describes how atomic orbitals combine to form molecular orbitals in molecules. For example, the formation of a hydrogen molecule (H_2) involves the overlap of two atomic orbitals to create a bonding molecular orbital.

$$H + H \longrightarrow H_2$$

1.1.9 Numerical Examples

Consider a particle with a given wavefunction $\psi(x)$. The probability density $|\psi(x)|^2$ represents the likelihood of finding the particle at a specific position x. Numerical integration can be employed to calculate the expectation value of position and momentum.

$$\langle x \rangle = \int_{-\infty}^{\infty} x |\psi(x)|^2 \, dx$$

$$\langle p \rangle = \int_{-\infty}^{\infty} \psi^*(x) \left(-i\hbar \frac{d}{dx} \right) \psi(x)\, dx$$

1.2   Historical Development of DFT

Density Functional Theory (DFT) has a rich historical background, marked by key milestones and the contributions of eminent physicists and chemists. The development of DFT can be traced back to the mid-20th century, with its roots in quantum mechanics and solid-state physics.

1.2.1   Early Concepts and Thomas-Fermi Model

The early development of DFT was influenced by the Thomas-Fermi model proposed by Llewellyn Thomas and Enrico Fermi in 1927. This model aimed to describe the electronic structure of atoms in terms of electron density. However, the model had limitations, and improvements were needed.

1.2.2   Hohenberg-Kohn Theorems

A significant breakthrough came in the 1960s with the formulation of the Hohenberg-Kohn theorems by Pierre Hohenberg and Walter Kohn. These theorems laid the foundation for DFT by establishing the one-to-one correspondence between the ground-state electron density and the external potential. Kohn later shared the Nobel Prize in Chemistry for this work in 1998.

1.2.3   Kohn-Sham DFT

In 1965, Walter Kohn and Lu Jeu Sham extended DFT by introducing the Kohn-Sham equations. This approach transformed the many-body problem of interacting electrons into a set of non-interacting electrons in an effective potential. The Kohn-Sham DFT is widely used in practical calculations due to its computational efficiency.

1.2.4 Exchange-Correlation Functionals

A critical aspect of DFT is the choice of exchange-correlation functional, which approximates the unknown exchange and correlation energies. Numerous functionals, such as the Local Density Approximation (LDA) and the Generalized Gradient Approximation (GGA), have been developed. Each functional has its strengths and limitations, and the choice depends on the system under investigation.

1.2.5 Advancements in Time-Dependent DFT (TDDFT)

In the 1990s, time-dependent DFT (TDDFT) emerged as an extension to study electronic excitations and dynamics. TDDFT has found applications in the investigation of optical properties, including electronic spectra and response to external fields.

1.2.6 Sample Equations

Let's consider a simplified form of the Kohn-Sham equation:

$$\left(-\frac{\hbar^2}{2m} \nabla^2 + V_{\text{eff}}(\mathbf{r}) \right) \psi_i(\mathbf{r}) = \epsilon_i \psi_i(\mathbf{r})$$

where V_{eff} is the effective potential, ψ_i is the Kohn-Sham orbital, and ϵ_i is the orbital energy.

1.2.7 Applications in Solid-State Physics

DFT has played a crucial role in understanding the electronic structure of solids. The prediction of material properties, such as band structures, has led to advancements in materials science. Examples include the study of semiconductors, insulators, and the discovery of novel materials with desirable electronic properties.

1.2.8 Numerical Examples

In the historical development of DFT, various numerical examples have played a crucial role in validating and advancing the theory. Let's explore a couple of key milestones:

Kohn-Sham Equations

The introduction of the Kohn-Sham equations by Walter Kohn and Pierre Hohenberg in 1964 was a groundbreaking development. These equations separate the many-body problem into non-interacting electrons in an effective potential, greatly simplifying the computational aspects.

$$\left(-\frac{\hbar^2}{2m} \nabla^2 + V_{\text{eff}}(\mathbf{r}) \right) \psi_i(\mathbf{r}) = \varepsilon_i \psi_i(\mathbf{r})$$

Numerical solutions of these equations have been pivotal in understanding the electronic structure of atoms and molecules.

LDA and GGA

The development of exchange-correlation functionals, such as the Local Density Approximation (LDA) and the Generalized Gradient Approximation (GGA), significantly improved the accuracy of DFT calculations.

$$E_{\text{XC}}[\rho(\mathbf{r})] = \int \rho(\mathbf{r}) \varepsilon_{\text{XC}}(\rho(\mathbf{r})) \, d\mathbf{r}$$

Numerical implementations of these functionals allowed researchers to tackle a wide range of systems, from small molecules to complex materials.

Quantum Espresso Code

The development and widespread use of computational packages, such as Quantum Espresso, have democratized access to DFT calculations. Researchers can perform numerically accurate simulations for diverse systems, including solids, surfaces, and nanostructures.

1.3 Basic Principles of DFT

Density Functional Theory (DFT) is founded on several fundamental principles that govern its application to the study of electronic structure and material properties. In this section, we will explore the core principles that underpin DFT and provide insights into its practical implementation.

1.3.1 The Hohenberg-Kohn Theorems

Density Functional Theory (DFT) is founded on the Hohenberg-Kohn theorems, which are fundamental principles governing the relationship between the external potential and the electron density of a quantum system.

First Hohenberg-Kohn Theorem

The first Hohenberg-Kohn theorem, proposed by Pierre Hohenberg and Walter Kohn in 1964, states that the external potential $V_{\text{ext}}(\mathbf{r})$ uniquely determines the ground-state electron density $\rho_0(\mathbf{r})$ of a many-electron system. Mathematically, this is expressed as:

$$V_{\text{ext}}(\mathbf{r}) \leftrightarrow \rho_0(\mathbf{r})$$

This theorem implies that all ground-state properties of a system can be derived from its electron density.

Second Hohenberg-Kohn Theorem

The second Hohenberg-Kohn theorem establishes the existence of a universal functional $F[\rho]$ that depends only on the electron density $\rho(\mathbf{r})$. The ground-state energy E_0 is minimized by the true electron density $\rho_0(\mathbf{r})$:

$$E_0 = \min_{\rho(\mathbf{r})} F[\rho]$$

This universal functional encapsulates all information about the system and is a key concept in the development of DFT.

The Hohenberg-Kohn theorems laid the foundation for the subsequent formulation of the Kohn-Sham equations, providing a rigorous framework for the practical application of DFT.

1.4   Fundamentals of Density Functional Theory (DFT)

1.4.1   The Kohn-Sham Equations

The Kohn-Sham equations are a set of mathematical equations derived by Walter Kohn and Lu Jeu Sham in 1965, extending the Hohenberg-Kohn theorems to make practical calculations feasible.

Kohn-Sham Single-Particle Equations

The Kohn-Sham equations introduce a set of auxiliary non-interacting electrons with fictitious potentials. The single-particle Kohn-Sham equations are given by:

$$\left[-\frac{\hbar^2}{2m}\nabla^2 + V_{\text{eff}}(\mathbf{r}) \right] \psi_i(\mathbf{r}) = \varepsilon_i \psi_i(\mathbf{r})$$

where $\psi_i(\mathbf{r})$ are the Kohn-Sham orbitals, ε_i are their corresponding eigenvalues, m is the electron mass, and $V_{\text{eff}}(\mathbf{r})$ is the effective potential.

Self-Consistent Iterative Scheme

The Kohn-Sham equations are solved self-consistently. The effective potential $V_{\text{eff}}(\mathbf{r})$ depends on the electron density, and the electron density is constructed from the occupied Kohn-Sham orbitals. The procedure is repeated until self-consistency is achieved.

The Kohn-Sham method provides a practical way to approach many-body quantum systems by transforming the problem into a set of single-particle equations, making it computationally feasible for a wide range of materials and systems.

1.4.2 Exchange-Correlation Functional

The exchange-correlation functional is a key component in the Kohn-Sham formulation of density functional theory (DFT). It combines the effects of exchange and correlation interactions among electrons.

Exchange Interaction

The exchange term accounts for the antisymmetrization requirement of the many-electron wavefunction. In the Kohn-Sham equations, the exchange term is expressed as:

$$E_{\text{exchange}} = -C_{\text{exchange}} \int \frac{\rho(\mathbf{r})\rho(\mathbf{r}')}{|\mathbf{r} - \mathbf{r}'|} \, d\mathbf{r} \, d\mathbf{r}'$$

where C_{exchange} is a proportionality constant and $\rho(\mathbf{r})$ is the electron density.

Correlation Interaction

The correlation term accounts for the dynamical correlation effects among electrons. It is often approximated using various functionals. One common form is the LDA (Local Density Approximation) for the correlation energy:

$$E_{\text{correlation}} = \int \varepsilon_c(\rho(\mathbf{r}))\rho(\mathbf{r}) \, d\mathbf{r}$$

where $\varepsilon_c(\rho(\mathbf{r}))$ is the correlation energy per particle as a functional of the electron density.

Exchange-Correlation Functional in DFT

The total exchange-correlation energy in DFT is given by the sum of the exchange and correlation contributions:

$$E_{\mathrm{xc}}[\rho(\mathbf{r})] = E_{\mathrm{exchange}} + E_{\mathrm{correlation}}$$

The choice of the exchange-correlation functional significantly influences the accuracy of DFT calculations for various systems.

1.4.3   Numerical Implementation

Implementing density functional theory (DFT) numerically involves solving the Kohn-Sham equations for a given system. The following steps outline a typical numerical implementation:

Discretization of Space

In order to solve the Kohn-Sham equations on a computer, the continuous spatial coordinates must be discretized. Common approaches include using a grid-based representation or a basis set to express the wavefunctions.

Kohn-Sham Equations

The central equations in DFT are the Kohn-Sham equations:

$$\left(-\frac{\hbar^2}{2m}\nabla^2 + V_{\mathrm{ext}}(\mathbf{r}) + V_{\mathrm{H}}(\mathbf{r}) + V_{\mathrm{xc}}[\rho(\mathbf{r})] \right) \psi_i(\mathbf{r}) = \varepsilon_i \psi_i(\mathbf{r})$$

where $V_{\mathrm{ext}}(\mathbf{r})$ is the external potential, $V_{\mathrm{H}}(\mathbf{r})$ is the Hartree potential, and $V_{\mathrm{xc}}[\rho(\mathbf{r})]$ is the exchange-correlation potential.

Self-Consistent Iteration

DFT is a self-consistent method, meaning that the electron density $\rho(\mathbf{r})$ appearing in $V_{\mathrm{H}}(\mathbf{r})$ and $V_{\mathrm{xc}}[\rho(\mathbf{r})]$ depends on the wavefunctions, which in turn depend on the electron density. This requires an iterative procedure until self-consistency is achieved.

Choice of Basis Set

The choice of basis set is crucial in the numerical implementation. Common choices include plane waves, localized atomic orbitals, or numerical atomic orbitals. Each basis set has its advantages and disadvantages, and the choice often depends on the specific system under investigation.

Integration Techniques

Various numerical techniques are employed for integrating quantities over space, such as evaluating the electron density, potentials, and the kinetic energy. Common methods include grid-based integration and numerical quadrature.

Convergence Criteria

To ensure the accuracy of the results, convergence criteria for the self-consistent iteration process are defined. These criteria may include thresholds for changes in the electron density, total energy, or wavefunctions.

Software Packages

Several software packages, such as Quantum ESPRESSO, VASP, and GPAW, provide tools for numerically implementing DFT calculations. These packages often include pre-implemented algorithms for solving the Kohn-Sham equations and handling various aspects of the numerical implementation.

Parallelization

Given the computational demands of DFT calculations, parallelization techniques are often employed to distribute the workload across multiple processors or nodes, improving efficiency and reducing computation time.

The numerical implementation of DFT involves a careful balance between accuracy and computational efficiency, and researchers choose specific methods and algorithms based on the nature of the system under study and the available computational resources.

1.4.4   Sample Equations

Let's explore some fundamental equations commonly encountered in density functional theory (DFT). These equations provide the basis for understanding the electronic structure of materials.

Kohn-Sham Equations

The Kohn-Sham equations govern the behavior of non-interacting electrons in an external potential V_{ext} and are the cornerstone of DFT:

$$\left(-\frac{\hbar^2}{2m}\nabla^2 + V_{\text{ext}}(\mathbf{r}) + V_{\text{H}}(\mathbf{r}) + V_{\text{xc}}[\rho(\mathbf{r})] \right) \psi_i(\mathbf{r}) = \varepsilon_i \psi_i(\mathbf{r})$$

Here, ψ_i represents the Kohn-Sham orbitals, ε_i are the corresponding eigenvalues, V_{H} is the Hartree potential, and V_{xc} is the exchange-correlation potential.

Total Energy Expression

The total energy of a system in DFT is expressed as a functional of the electron density $\rho(\mathbf{r})$:

$$E[\rho] = T[\rho] + V_{\text{ext}}[\rho] + V_{\text{H}}[\rho] + E_{\text{xc}}[\rho]$$

where $T[\rho]$ is the kinetic energy, $V_{\text{ext}}[\rho]$ is the external potential energy, $V_{\text{H}}[\rho]$ is the Hartree energy, and $E_{\text{xc}}[\rho]$ is the exchange-correlation energy.

Density Functional

The key concept in DFT is the exchange-correlation functional $E_{\text{xc}}[\rho]$, which encapsulates the effects of electron-electron interactions. Common approximations include the Local Density Approximation (LDA) and the Generalized Gradient Approximation (GGA).

Hohenberg-Kohn Theorems

The Hohenberg-Kohn theorems establish the foundation of DFT, stating that the external potential V_{ext} uniquely determines the ground-state electron density and vice versa.

$$E_0 = \inf \left\{ E[\rho] : \rho(\mathbf{r}) \mapsto n(\mathbf{r}) \right\}$$

These sample equations showcase the fundamental principles and mathematical expressions that govern density functional theory, providing a starting point for understanding the theoretical framework.

1.4.5 Applications in Molecular Systems

Density Functional Theory (DFT) finds extensive applications in understanding and predicting various properties of molecular systems. In this section, we explore some key applications and highlight the versatility of DFT in molecular studies.

Electronic Structure of Molecules

One of the primary applications of DFT is in determining the electronic structure of molecules. The Kohn-Sham equations provide a framework to calculate molecular orbitals, energy levels, and electron density distributions. This information is crucial for understanding bonding patterns and electronic configurations in molecules.

Molecular Dynamics Simulations

DFT is employed in molecular dynamics simulations to study the time-dependent behavior of molecular systems. By integrating the equations of motion derived from the electronic structure, researchers can investigate molecular vibrations, rotations, and conformational changes over time.

Prediction of Spectroscopic Properties

DFT plays a pivotal role in predicting spectroscopic properties of molecules. From infrared (IR) and ultraviolet-visible (UV-Vis) spectra to nuclear magnetic resonance (NMR) parameters, DFT calculations provide insights into the experimentally observed spectral features, aiding in the interpretation of experimental data.

Chemical Reactivity and Reaction Mechanisms

Understanding chemical reactivity and reaction mechanisms is essential in chemistry. DFT allows researchers to explore reaction pathways, transition states, and reaction energetics. This information is valuable for predicting reaction outcomes and designing new synthetic routes.

Solvent Effects and Molecular Interactions

Incorporating solvent effects is crucial for accurate modeling of molecular systems in realistic environments. DFT can be extended to study solvent effects, providing insights into how molecular interactions change in different solvents and under varying conditions.

Design of Functional Materials

DFT is widely used in the design and optimization of functional materials. Whether it's catalysts for chemical reactions, organic semiconductors for electronic devices, or materials for energy storage, DFT calculations aid in predicting and optimizing the properties of these materials.

These applications illustrate the broad utility of DFT in unraveling the complexities of molecular systems and provide a foundation for advancing research in various fields.

1.4.6   Bonding Diagrams

Bonding diagrams are essential visual tools for illustrating the nature of chemical bonds within molecules. In this section, we explore the construction and interpretation of bonding diagrams using Density Functional Theory (DFT).

Lewis Structures

Lewis structures, a fundamental tool in chemistry, depict the arrangement of atoms and valence electrons in a molecule. DFT provides a theoretical framework for generating Lewis structures by optimizing the electronic configuration to minimize the total energy of the system.

$$
\begin{array}{c}
H \\
| \\
H - C \\
| \diagdown H \\
H
\end{array}
$$

Figure 1.1: Lewis structure of methane (CH_4) obtained through DFT calculations.

Molecular Orbital Diagrams

Molecular orbital (MO) diagrams offer a more detailed view of the distribution of electrons in a molecule. DFT calculations can predict molecular orbitals and their energy levels, enabling the construction of MO diagrams for molecules of interest.

H2   O2

Figure 1.2: Molecular orbital diagrams for diatomic molecules (H_2 and O_2) based on DFT calculations.

Resonance Structures

Resonance is a concept that arises in molecules with delocalized electrons. DFT can be employed to explore the different resonance structures of a molecule and assess their contribution to the overall electronic structure.

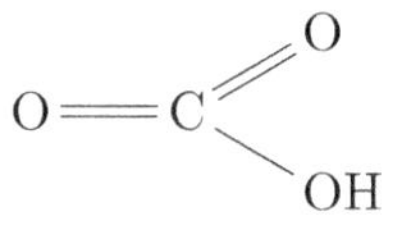

$$O = C \underset{OH}{\overset{O}{<}}$$

Figure 1.3: Resonance structures of the carbonate ion $(CO_3{}^{2-})$ obtained using DFT calculations.

Bonding in Nanomaterials

DFT is particularly powerful in studying bonding in nanomaterials, such as graphene and carbon nanotubes. The unique bonding patterns and electronic structures of these materials can be elucidated through DFT calculations.

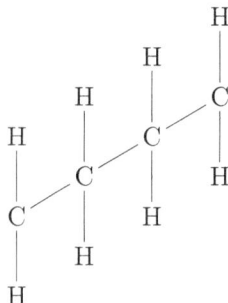

Figure 1.4: DFT-based bonding diagram of a carbon nanotube.

These examples demonstrate how DFT serves as a valuable tool for constructing bonding diagrams and gaining deeper insights into the nature of chemical bonds in a variety of molecular systems.

1.4.7   Advancements and Challenges

As Density Functional Theory (DFT) continues to evolve, several advancements and challenges shape the landscape of its applications and development. In this subsection, we explore some of the notable advancements and ongoing challenges in the field.

Advancements in Methodology

Recent years have witnessed significant advancements in DFT methodologies. Improved exchange-correlation functionals, such as meta-GGA and hybrid functionals, have enhanced the accuracy of DFT calculations. Additionally, developments in time-dependent DFT (TDDFT) enable the study of excited states and electronic transitions.

High-Performance Computing

The increasing availability of high-performance computing resources has revolutionized the scale and complexity of DFT calculations. Researchers can now tackle larger and more intricate systems, ranging from complex biomolecules to materials with diverse properties.

Machine Learning Integration

The integration of machine learning techniques with DFT has emerged as a promising avenue. Machine learning models can accelerate materials discovery, predict properties, and optimize DFT calculations, making the computational exploration of vast chemical spaces more efficient.

Challenges in Accuracy

Despite advancements, achieving high accuracy in DFT calculations remains a challenge. Certain systems, especially those involving strong correlation effects or transition metal complexes, may require sophisticated methodologies beyond the capabilities of standard DFT.

Treatment of Strongly Correlated Systems

Understanding and accurately describing strongly correlated systems pose a significant challenge for DFT. These systems often exhibit complex electronic structures, and developing suitable functionals to capture their behavior accurately is an ongoing area of research.

Relativistic Effects

Incorporating relativistic effects into DFT calculations is another frontier. Systems involving heavy elements or high-speed processes require accurate treatment of relativistic effects, demanding the development of specialized functionals and methodologies.

Interdisciplinary Applications

DFT is increasingly finding applications beyond traditional chemistry and materials science. Interdisciplinary research areas, including biology, environmental science, and catalysis, present new challenges and opportunities for DFT methodologies.

Navigating these advancements and challenges is crucial for researchers and practitioners in the field of Density Functional Theory, fostering innovation and broadening the scope of applications.

Chapter 2

Fundamentals of Quantum Chemistry

2.1 Atomic Structure and Electronic Configuration

Understanding atomic structure and electronic configuration is crucial for grasping the principles of quantum chemistry. In this section, we explore these fundamental concepts with real examples.

2.1.1 Atomic Structure

The Bohr model, though simplistic, introduces key ideas. Consider the hydrogen atom (H). It consists of a single proton in the nucleus and one electron in orbit. The Bohr model representation for hydrogen is given by:

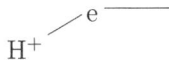

This representation illustrates the nucleus (proton) and the electron in a specific orbit.

2.1.2   Quantum Mechanical Model

The quantum mechanical model, also known as the quantum model or wave-mechanical model, serves as the foundation for understanding the behavior of electrons within an atom. This model builds upon the principles of quantum mechanics to describe the distribution of electrons in atomic orbitals.

Wave-Particle Duality

One of the fundamental tenets of the quantum mechanical model is the wave-particle duality, which asserts that particles, including electrons, exhibit both wave-like and particle-like properties. This duality is encapsulated in the concept of wavefunctions, mathematical functions that describe the probability density of finding an electron at a particular location.

Schrödinger Equation

The Schrödinger equation lies at the heart of the quantum mechanical model. It is a partial differential equation that describes how the wavefunction of a physical system changes over time. For an electron in an atom, the Schrödinger equation takes into account the potential energy due to the nucleus and the kinetic energy of the electron.

$$\hat{H}\Psi = E\Psi \tag{2.1}$$

where $\hat{H}$ is the Hamiltonian operator, Ψ is the wavefunction, and E is the total energy of the system.

Quantum Numbers

The quantum numbers are integral to the quantum mechanical model as they provide a unique set of identifiers for each electron in an atom. These include:

- **Principal Quantum Number** (n): Represents the energy level of an electron.

- **Angular Momentum Quantum Number** (l): Determines the shape of the orbital.

- **Magnetic Quantum Number** (m_l): Specifies the orientation of the orbital in space.

- **Spin Quantum Number** (m_s): Describes the intrinsic spin of the electron.

Atomic Orbitals

In the quantum mechanical model, atomic orbitals are mathematical functions that describe the regions of high probability for finding electrons. The most common orbitals include the s, p, d, and f orbitals, each with distinct shapes and orientations.

Electron Configuration

The arrangement of electrons in an atom is described by its electron configuration, which specifies the distribution of electrons among different atomic orbitals. Understanding electron configurations is crucial for predicting the chemical behavior of elements.

The quantum mechanical model revolutionized our understanding of atomic structure, providing a powerful framework for predicting and interpreting the behavior of electrons in atoms.

2.1.3 Electronic Configuration

The electronic configuration of an atom is a crucial aspect of its quantum mechanical description. It provides information about how electrons are distributed among different atomic orbitals, governed by the principles of quantum mechanics.

Notation and Rules

The electronic configuration is typically represented using a series of numbers and letters. The following rules and notations are commonly employed:

- **Aufbau Principle**: Electrons fill the lowest energy orbitals first before moving to higher energy ones.

- **Pauli Exclusion Principle**: No two electrons in an atom can have the same set of quantum numbers, including spin.

- **Hund's Rule**: Electrons occupy orbitals singly before pairing up.

Representation

The electronic configuration is often represented using the notation:

$$1s^2\, 2s^2\, 2p^6\, 3s^2\, 3p^6\, 4s^2\, 3d^{10}\, 4p^6\, 5s^2\, 4d^{10}\, 5p^6\, \ldots \qquad (2.2)$$

Here, each term represents the occupancy of a specific atomic orbital, with the principal quantum number (n), the azimuthal quantum number (l), and the number of electrons in that orbital.

Examples

Hydrogen (H): $1s^1$

Oxygen (O): $1s^2\, 2s^2\, 2p^4$

Iron (Fe): $1s^2\, 2s^2\, 2p^6\, 3s^2\, 3p^6\, 4s^2\, 3d^6$

Let's explore the electronic configuration of carbon (C). With $Z = 6$, the electronic configuration is $1s^2 2s^2 2p^2$. The Lewis structure for methane (CH_4) can be represented as follows:

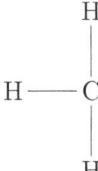

This representation illustrates the tetrahedral arrangement of hydrogen atoms around the carbon atom.

Significance

Understanding the electronic configuration is crucial for predicting the chemical properties and reactivity of elements. It provides insights into the arrangement of electrons in atoms, guiding the interpretation of the periodic table and chemical bonding.

The electronic configuration serves as a foundation for more advanced topics in quantum chemistry and density functional theory, playing a central role in the description of atomic and molecular properties.

2.2 Molecular Orbital Theory

Molecular Orbital Theory (MOT) is a fundamental concept in quantum chemistry that provides a theoretical framework for understanding the electronic structure of molecules. It extends the ideas of atomic orbitals to describe the distribution of electrons in molecules and predict their properties.

2.2.1 Basic Principles

Molecular Orbital Theory is based on the following key principles:

- **Linear Combination of Atomic Orbitals (LCAO)**: Molecular orbitals are formed by linear combinations of atomic orbitals from the participating atoms in a molecule.

- **Superposition Principle**: The molecular orbital is a superposition of the atomic orbitals, and its shape and energy depend on the contributions from individual atomic orbitals.

- **Wave Function Symmetry**: Molecular orbitals can have different symmetries, and their shapes are determined by the constructive or destructive interference of atomic orbitals.

2.2.2   Formation of Molecular Orbitals

The combination of atomic orbitals to form molecular orbitals involves the interaction of atomic orbitals from different atoms. The most common types of molecular orbitals include:

- **Sigma (σ) Molecular Orbitals**: Formed by head-to-head or tail-to-tail overlap of atomic orbitals along the internuclear axis.

- **Pi (π) Molecular Orbitals**: Formed by the side-to-side overlap of atomic orbitals above and below the internuclear axis.

- **Delta (δ) Molecular Orbitals**: Involve the overlap of d orbitals in more complex molecules.

2.2.3   Application in Diatomic Molecules

Molecular Orbital Theory has been successfully applied to diatomic molecules, predicting their electronic structure and properties. Examples include:

Hydrogen Molecule (H_2):   The molecular orbital diagram for H_2 involves the combination of two hydrogen atomic orbitals to form a bonding σ orbital and an antibonding σ^* orbital.

Oxygen Molecule (O_2): O_2 involves the combination of atomic orbitals from two oxygen atoms to form σ and π molecular orbitals, contributing to the stability of the molecule.

2.2.4 Advancements and Applications

Molecular Orbital Theory serves as a foundation for understanding the electronic structure of complex molecules, predicting bond strengths, and explaining spectroscopic properties. It has found applications in various fields, including computational chemistry, materials science, and drug design.

Understanding the principles of Molecular Orbital Theory is essential for a comprehensive grasp of quantum chemistry and its applications in diverse areas of research and technology.

2.3 Chemical Bonding

Chemical bonding is a fundamental concept in chemistry that describes the interactions between atoms to form molecules. These interactions are crucial in understanding the properties and behaviors of matter. In this section, we will explore various types of chemical bonds and their significance.

2.3.1 Covalent Bonds

Covalent bonds involve the sharing of electrons between atoms. In a covalent bond, atoms contribute electrons to create a shared electron pair. For example, the formation of a hydrogen molecule (H_2) involves two hydrogen atoms sharing a pair of electrons:

$$H \diagup^{H}$$

The line between the hydrogen atoms represents the shared electron pair.

2.3.2 Ionic Bonds

Ionic bonds result from the transfer of electrons from one atom to another. This leads to the formation of ions with opposite charges that attract each other. An example is the bond between sodium (Na) and chlorine (Cl) in sodium chloride (NaCl):

$$Na^+ + Cl \longrightarrow \longrightarrow NaCl$$

2.3.3   Metallic Bonds

Metallic bonds occur between metal atoms and involve a sea of delocalized electrons. This type of bonding is responsible for the unique properties of metals, such as conductivity and malleability.

2.3.4   Hydrogen Bonds

Hydrogen bonds are special interactions between a hydrogen atom bonded to a highly electronegative atom (usually oxygen, nitrogen, or fluorine) and another electronegative atom. These bonds are crucial in the structure of molecules like water (H_2O).

2.3.5   Van der Waals Forces

Van der Waals forces are weak intermolecular forces that arise due to temporary fluctuations in electron distribution. These forces include London dispersion forces, dipole-dipole interactions, and hydrogen bonding.

Understanding these various types of chemical bonds is essential for predicting molecular structures and properties.

Chapter 3

Mathematics and
Formalism in DFT

3.1 Mathematical Basis of DFT

Density Functional Theory (DFT) is a powerful theoretical framework for describing electronic structure and properties of materials. At its core, DFT relies on the Hohenberg-Kohn theorems, which establish that the ground-state electron density uniquely determines the system's electronic and external potentials. This section provides a mathematical foundation for understanding key concepts in DFT.

3.1.1 Hohenberg-Kohn Theorems

The Hohenberg-Kohn theorems state:

1. The ground-state electron density $n(\mathbf{r})$ uniquely determines the external potential $V_{\text{ext}}(\mathbf{r})$ up to an additive constant.

2. The ground-state energy E_0 is a unique functional of the electron density, given by:

$$E_0[n] = T_s[n] + \int V_{\text{ext}}(\mathbf{r}) n(\mathbf{r}) \, d\mathbf{r} + \frac{1}{2} \iint \frac{n(\mathbf{r}) n(\mathbf{r}')}{|\mathbf{r} - \mathbf{r}'|} \, d\mathbf{r} \, d\mathbf{r}' + E_{\text{xc}}[n] + E_{\text{ion}}$$

Here, $T_s[n]$ is the kinetic energy of a non-interacting electron gas, and $E_{\text{xc}}[n]$ is the exchange-correlation energy functional.

3.1.2   Kohn-Sham Equations

To simplify the many-body problem, the Kohn-Sham approach introduces a set of fictitious non-interacting electrons with the same electron density. The Kohn-Sham equations are given by:

$$\left(-\frac{\hbar^2}{2m} \nabla^2 + V_{\text{eff}}(\mathbf{r}) \right) \phi_i(\mathbf{r}) = \varepsilon_i \phi_i(\mathbf{r})$$

$$n(\mathbf{r}) = \sum_i^N |\phi_i(\mathbf{r})|^2$$

Here, $V_{\text{eff}}(\mathbf{r}) = V_{\text{ext}}(\mathbf{r}) + V_{\text{H}}(\mathbf{r}) + V_{\text{xc}}(\mathbf{r})$ is the effective potential, including external, Hartree, and exchange-correlation contributions.

3.1.3   Example: Hydrogen Atom

Consider a hydrogen atom as an example. The electron density $n(\mathbf{r})$ for the ground state can be determined, and the corresponding Kohn-Sham equations solved to obtain the energy levels and wavefunctions.

3.1.4   Numerical Implementation

Numerical methods such as the finite difference or plane wave basis set approaches are commonly employed to solve the Kohn-Sham equations. These methods discretize the continuous problem, making it amenable to computational solutions.

3.2 Formalism and Equations in DFT

Density Functional Theory (DFT) is a powerful tool for understanding the electronic structure of materials. This section explores the formalism and key equations that govern DFT calculations.

3.2.1 Kohn-Sham Equations

At the heart of DFT lies the Kohn-Sham equations, which provide a framework for simplifying the many-body problem. The Kohn-Sham equations are as follows:

$$\left(-\frac{\hbar^2}{2m}\nabla^2 + V_{\text{eff}}(\mathbf{r})\right)\phi_i(\mathbf{r}) = \varepsilon_i\phi_i(\mathbf{r})$$

$$n(\mathbf{r}) = \sum_i^N |\phi_i(\mathbf{r})|^2$$

Here, $V_{\text{eff}}(\mathbf{r}) = V_{\text{ext}}(\mathbf{r}) + V_{\text{H}}(\mathbf{r}) + V_{\text{xc}}(\mathbf{r})$ is the effective potential, including external, Hartree, and exchange-correlation contributions.

3.2.2 Exchange-Correlation Functional

The exchange-correlation functional $E_{\text{xc}}[n]$ captures the quantum mechanical effects beyond classical electrostatics. Common approximations include the Local Density Approximation (LDA) and the Generalized Gradient Approximation (GGA).

3.2.3 Example: Carbon Atom

Consider a simple example with a carbon atom. The Kohn-Sham equations for the ground state can be solved numerically, and the resulting electron density and energy levels provide insights into the atom's electronic structure.

3.2.4   Chemical Equations

Incorporating chemical equations is essential for practical applications of DFT. For example, the chemical reaction for the formation of water (H_2O) can be represented as:

$$2\,H_2(g) + O_2(g) \longrightarrow 2\,H_2O(g)$$

3.2.5   Bonding Diagrams

Bonding diagrams are valuable tools for visualizing electron distribution. For instance, the Lewis structure of methane (CH_4) can be represented as:

$$
\begin{array}{ccc}
H & & H \\
 & H & \\
H & & C
\end{array}
$$

3.2.6   Numerical Implementation

Numerical methods such as the finite element method or plane wave basis set are employed for solving the Kohn-Sham equations. Implementing these methods requires careful consideration of convergence criteria and computational resources.

3.3   Approximations in DFT

While Density Functional Theory (DFT) provides a robust framework for studying electronic structure, practical implementations often involve approximations to make calculations computationally feasible. This section explores key approximations used in DFT.

3.3.1 Local Density Approximation (LDA)

One of the earliest and simplest approximations is the Local Density Approximation (LDA). It assumes that the exchange-correlation energy per electron is uniform throughout space, depending only on the local electron density. The LDA exchange-correlation functional is given by:

$$E_{\text{xc}}^{\text{LDA}}[n] = \int n(\mathbf{r}) \varepsilon_{\text{xc}}^{\text{LDA}}(n(\mathbf{r})) \, d\mathbf{r}$$

Here, $\varepsilon_{\text{xc}}^{\text{LDA}}(n(\mathbf{r}))$ is the LDA exchange-correlation energy density.

3.3.2 Generalized Gradient Approximation (GGA)

Building upon LDA, the Generalized Gradient Approximation (GGA) includes information about the gradient of the electron density. It provides a more accurate description of the exchange-correlation energy. The GGA functional is expressed as:

$$E_{\text{xc}}^{\text{GGA}}[n] = \int n(\mathbf{r}) \varepsilon_{\text{xc}}^{\text{GGA}}(n(\mathbf{r}), \nabla n(\mathbf{r})) \, d\mathbf{r}$$

GGA functionals are widely used in modern DFT calculations.

3.3.3 Hybrid Functionals

Hybrid functionals are a class of exchange-correlation functionals used in density functional theory (DFT) that combine the advantages of both local and non-local functionals. These functionals aim to provide more accurate predictions of electronic structure and properties compared to standard DFT functionals.

Basic Concept

The key idea behind hybrid functionals is the combination of a fraction of Hartree-Fock exchange with exchange-correlation functionals commonly used in DFT, such as the local density approximation (LDA) or generalized gradient approximation (GGA). This hybridization helps address the limitations of standard functionals in describing certain types of electronic interactions, especially those involving dispersion forces.

Mathematical Formulation

A popular form of hybrid functional is the Becke, 3-parameter, Lee-Yang-Parr (B3LYP) functional, which combines Becke's three-parameter exchange functional with the correlation functional of Lee, Yang, and Parr. The mathematical formulation of B3LYP can be expressed as:

$$E_{xc}^{B3LYP} = a \cdot E_X^{HF} + (1 - a) \cdot E_X^{DFT} + b \cdot E_C^{LDA} + (1 - b) \cdot E_C^{GGA}$$

3.3.4   Example: Water Molecule

To illustrate the application of hybrid functionals in practical calculations, let's consider the water molecule (H_2O). Water is a fundamental molecule with interesting electronic properties, making it a suitable example to showcase the capabilities of hybrid functionals.

Geometry Optimization

First, we perform a geometry optimization of the water molecule using a hybrid functional, such as B3LYP. The goal is to find the most stable configuration of the molecule by minimizing its total energy. The optimized geometry can provide insights into bond lengths and angles, which are crucial for understanding the molecular structure.

$$\text{Initial Geometry:} \quad H-O-H$$
$$\text{Optimized Geometry:} \quad H-O-H$$

Electronic Structure

Next, we analyze the electronic structure of the water molecule using the hybrid functional. This involves studying the distribution of electron density, molecular orbitals, and the energy levels of the molecular system. Hybrid functionals are known for providing more accurate descriptions of electronic interactions, making them suitable for predicting electronic properties.

Energetics

Hybrid functionals allow for a more accurate prediction of energetic properties. We can calculate the molecular energy, including the electronic energy, vibrational contributions, and other relevant terms. This information is crucial for understanding the stability and reactivity of the water molecule in different environments.

Comparison with Standard Functionals

To highlight the impact of using a hybrid functional, we can compare the results obtained with B3LYP to those obtained with standard functionals, such as LDA or GGA. This comparison will showcase the improvements brought by the hybrid functional in terms of accuracy and reliability.

In summary, studying the water molecule with hybrid functionals provides a practical example of how these functionals enhance the accuracy of electronic structure calculations. Researchers often use such examples to validate and benchmark the performance of hybrid functionals in different molecular systems.

3.3.5 Accuracy-Computational Cost Trade-off

Choosing an appropriate level of approximation involves a trade-off between accuracy and computational cost. Researchers need to select the most suitable approximation based on the system under investigation and the available computational resources.

3.3.6 Chemical Reactions and Bonding

DFT approximations play a crucial role in studying chemical reactions and bonding. For instance, the calculation of activation energies and reaction pathways relies on accurate exchange-correlation functionals.

3.3.7 Limitations and Future Developments

While approximations make DFT computationally viable, they also introduce limitations. Ongoing research focuses on developing more accurate functionals and improving the trade-off between accuracy and computational cost.

Chapter 4

Computational Methods in DFT

4.1 Numerical Techniques

Numerical techniques play a crucial role in solving the mathematical equations that arise in Density Functional Theory (DFT). This section explores some of the key numerical methods used in DFT calculations.

4.1.1 Discretization of Space

To perform DFT calculations, continuous space must be discretized. This involves dividing the system into a grid or mesh. The discretization of space is essential for solving the Kohn-Sham equations numerically.

4.1.2 Grid-Based Methods

One common numerical technique is grid-based methods, where the discretized space forms a grid. The electron density, potential, and other quantities are evaluated at grid points. This approach is efficient for systems with periodic boundary conditions.

4.1.3   Finite Difference Methods

Finite Difference Methods (FDM) play a crucial role in solving partial differential equations (PDEs) numerically. In the context of density functional theory (DFT), FDM is commonly employed for solving the Kohn-Sham equations and obtaining electronic structures. Here, we delve into the fundamentals of FDM and its application in DFT calculations.

Basic Principles

At the core of FDM lies the discretization of continuous differential equations into a finite set of algebraic equations. In the context of DFT, the Kohn-Sham equations are typically discretized in space and solved iteratively. The basic principles involve approximating derivatives using finite differences and formulating a system of linear equations.

Discretization Schemes

Various discretization schemes are employed based on the specific differential equations being solved. Common schemes include central differences, forward differences, and backward differences. The choice of scheme depends on factors such as stability, accuracy, and computational efficiency.

Spatial Grids

In DFT, the spatial domain is discretized into a grid, and the electronic wavefunctions and potentials are evaluated at discrete grid points. The selection of an appropriate spatial grid has a significant impact on the accuracy of the calculations. Fine grids provide higher accuracy but demand more computational resources.

Time Evolution Methods

In time-dependent DFT (TDDFT), finite difference methods are used to propagate the electronic density over time. Time evolution methods, such as the

Crank-Nicolson scheme, enable the simulation of dynamic processes, such as electronic excitations.

Implementation in Software

Numerical libraries and software packages for quantum chemistry and DFT often incorporate FDM for solving electronic structure problems. Understanding the implementation details in popular software tools is essential for researchers conducting DFT simulations.

Challenges and Considerations

While FDM is a powerful tool, it comes with challenges such as numerical instabilities and convergence issues. Researchers must be aware of these challenges and adopt appropriate strategies to ensure reliable and accurate results.

In summary, Finite Difference Methods form a cornerstone in the numerical solution of DFT problems, providing a versatile and widely used approach for obtaining electronic structures and properties.

4.1.4 Fast Fourier Transform (FFT)

The Fast Fourier Transform (FFT) is a crucial algorithm in computational mathematics and signal processing. In the context of density functional theory (DFT), FFT is commonly employed for efficient numerical solutions. Let's explore the fundamentals and mathematical equations associated with FFT.

Basic Concepts

FFT is an algorithm to compute the discrete Fourier transform and its inverse efficiently. It exploits the symmetry properties of the Fourier transform to reduce the computational complexity from $O(N^2)$ to $O(N \log N)$, where N is the number of data points.

Discrete Fourier Transform (DFT)

The Discrete Fourier Transform of a sequence x_n is defined as:

$$X_k = \sum_{n=0}^{N-1} x_n e^{-\frac{2\pi i}{N} kn}$$

where X_k is the transformed sequence and N is the total number of data points.

FFT Algorithm

The FFT algorithm recursively divides the DFT computation into smaller sub-problems. The Cooley-Tukey radix-2 algorithm is a widely used FFT variant. For a sequence of length $N = 2^p$, the FFT can be expressed as:

$$X_k = \sum_{m=0}^{N/2-1} x_{2m} e^{-\frac{2\pi i}{N} km} + e^{-\frac{2\pi i}{N} k} \sum_{m=0}^{N/2-1} x_{2m+1} e^{-\frac{2\pi i}{N} km}$$

Implementation in DFT

In DFT calculations, FFT is employed to efficiently compute convolutions and other operations in reciprocal space. It significantly speeds up the computation of electronic densities and potentials, making large-scale DFT simulations feasible.

Computational Considerations

While FFT greatly accelerates computations, proper consideration must be given to factors such as grid size, grid spacing, and parallelization strategies. Optimal FFT implementations are crucial for achieving high-performance computing in DFT simulations.

In summary, Fast Fourier Transform is a fundamental algorithm with wide applications in DFT, enabling efficient and scalable numerical solutions for electronic structure calculations.

4.1.5 Pseudopotentials

The use of pseudopotentials is another numerical technique to reduce computational complexity. Pseudopotentials replace the core electrons in the calculation, allowing for a more efficient representation of the electronic structure.

4.1.6 Example: Carbon Nanotube

The electronic structure of a carbon nanotube (C_{60}) can be represented by illustrating its molecular orbitals. Here's a simplified diagram:

$$C = C - C - C - C - C$$

This representation shows the cyclic structure of C_{60}, and each carbon atom contributes to the overall electronic structure.

4.1.7 Convergence and Accuracy

Achieving convergence in DFT calculations is essential for obtaining reliable results. Researchers must carefully select parameters such as the grid spacing, energy cutoff, and convergence criteria to ensure accurate and stable calculations.

4.1.8 Parallelization

Given the computational intensity of DFT calculations, parallelization is often employed to distribute the workload across multiple processors or nodes. Parallel numerical algorithms enhance the efficiency of large-scale DFT simulations.

4.1.9 Computational Cost Considerations

While advances in numerical techniques and computing power have expanded the scope of DFT calculations, researchers must still consider the computational cost. Balancing accuracy and computational efficiency is a critical aspect of numerical DFT simulations.

4.1.10   Applications in Materials Science

Numerical techniques in DFT find extensive applications in materials science, allowing researchers to explore the electronic properties, stability, and reactivity of various materials.

4.2   Basis Sets and Pseudopotentials

Basis sets and pseudopotentials are integral components of Density Functional Theory (DFT) calculations. In this section, we explore their roles and significance in accurately describing the electronic structure of materials.

4.2.1   Basis Sets

Basis sets are sets of functions used to represent the electronic wavefunctions in a DFT calculation. They are crucial for approximating the behavior of electrons in a material. Common types of basis sets include:

Localized Basis Sets

Localized basis functions are often used to represent molecular orbitals. Here's an example of a localized basis set for a simple molecule:

In this molecule, we have three localized basis functions associated with each hydrogen atom:

- Hydrogen 1: ϕ_1 (localized at the first hydrogen atom)

- Hydrogen 2: ϕ_2 (localized at the second hydrogen atom)

- Carbon: ϕ_3 (localized at the carbon atom)

The molecular orbital can be represented as a linear combination of these localized basis functions:

$$\psi = c_1\phi_1 + c_2\phi_2 + c_3\phi_3$$

where c_1, c_2, c_3 are coefficients determined by the quantum chemical calculations.

Plane-Wave Basis Sets

4.3 Plane-Wave Basis Sets

In density functional theory for periodic systems, plane-wave basis sets are often employed. The electronic wavefunctions are expanded as a sum of plane waves:

$$\psi(r) = \sum_{\mathbf{G}} c_{\mathbf{G}} e^{i\mathbf{G}\cdot\mathbf{r}}$$

Here, $\psi(r)$ is the electronic wavefunction, $\mathbf{G}$ are reciprocal lattice vectors, $\mathbf{r}$ is the position vector, and $c_{\mathbf{G}}$ are expansion coefficients.

The kinetic energy operator in a periodic system can be expressed in terms of the plane-wave basis:

$$\hat{T} = -\frac{1}{2}\sum_{\mathbf{G}} \nabla^2 e^{i\mathbf{G}\cdot\mathbf{r}}$$

Additionally, the electron-electron interaction term can be written using the reciprocal lattice vectors:

$$\hat{V}_{ee} = \frac{1}{2}\sum_{\mathbf{G}_1,\mathbf{G}_2,\mathbf{G}_3} \frac{4\pi}{|\mathbf{G}_1+\mathbf{G}_2|^2} c^*_{\mathbf{G}_1} c^*_{\mathbf{G}_2} c_{\mathbf{G}_3} c_{\mathbf{G}_4} \delta_{\mathbf{G}_1+\mathbf{G}_2,\mathbf{G}_3+\mathbf{G}_4}$$

In these expressions, ∇^2 is the Laplacian operator, and $\delta_{\mathbf{G}_1+\mathbf{G}_2,\mathbf{G}_3+\mathbf{G}_4}$ is the Kronecker delta.

Plane-wave basis sets are particularly suitable for treating periodic systems, such as crystals and surfaces.

4.3.1   Pseudopotentials

Pseudopotentials are used to approximate the interaction between valence electrons and atomic cores. They replace the core electrons with effective potentials, reducing the computational cost while maintaining accuracy. Common types include:

Norm-Conserving Pseudopotentials

Norm-conserving pseudopotentials preserve the norm of the wavefunction, ensuring accurate representation of the electronic states. They are widely employed in DFT calculations for their reliability.

Ultrasoft Pseudopotentials

4.4   Ultrasoft Pseudopotentials

Ultrasoft pseudopotentials are widely used in density functional theory calculations to describe the electron-ion interactions in a computationally efficient manner. They are constructed to have smooth radial shapes, allowing for a lower plane-wave cutoff in the electronic wavefunction expansion.

The ultrasoft pseudopotential $\hat{V}_{\text{US}}$ typically consists of a nonlocal and a local part:

$$\hat{V}_{\text{US}} = \hat{V}_{\text{US,nonlocal}} + \hat{V}_{\text{US,local}}$$

The nonlocal part is given by a sum over angular momentum channels:

$$\hat{V}_{\text{US,nonlocal}} = \sum_{lm} \int \mathrm{d}r \, r^2 v_{lm}(r) \, Y_{lm}(\theta, \phi) \, n(\mathbf{r})$$

Here, $v_{lm}(r)$ are the nonlocal pseudopotential functions, $Y_{lm}(\theta, \phi)$ are spherical harmonics, and $n(\mathbf{r})$ is the electron charge density.

The local part is typically written as:

$$\hat{V}_{\text{US,local}} = \sum_i \alpha_i n_i(\mathbf{r}) + \sum_i \beta_i \nabla^2 n_i(\mathbf{r})$$

Here, α_i and β_i are parameters, and $n_i(\mathbf{r})$ are smooth functions of the electron charge density.

Ultrasoft pseudopotentials play a crucial role in reducing the computational cost of DFT calculations for complex systems.

4.4.1 Choosing Basis Sets and Pseudopotentials

4.5 Choosing Basis Sets and Pseudopotentials

In DFT calculations, the choice of basis sets and pseudopotentials is crucial for obtaining accurate and efficient results. The basis set represents the set of functions used to describe the electronic wavefunctions, while pseudopotentials are employed to approximate the interaction between electrons and atomic nuclei.

4.5.1 Basis Sets

The choice of basis sets depends on the system under investigation. Gaussian basis sets, such as the popular Pople basis sets (e.g., 6-31G(d)), are commonly used for molecular systems. For periodic systems, plane-wave basis sets are preferred due to their efficiency in handling periodic boundary conditions.

Molecular System: 6-31G(d) Basis Set

Periodic System: Plane-Wave Basis Set

Additionally, basis sets with polarization and diffuse functions may be necessary to accurately describe electronic correlation and molecular properties.

4.5.2 Pseudopotentials

Pseudopotentials are used to replace the core electrons of atoms, allowing for a more efficient description of the valence electrons. The choice of pseudopotentials should be compatible with the chosen basis set.

Ultrasoft Pseudopotentials for Efficient Plane-Wave Basis Sets

For accurate results, it's essential to validate the chosen combination of basis sets and pseudopotentials against experimental data or higher-level theoretical calculations.

4.5.3 Example: Silicon Crystal

Let's consider the example of a silicon crystal, a common material in condensed matter physics. Silicon has a diamond cubic crystal structure, and its electronic properties are of interest for various applications.

For silicon crystal calculations, a plane-wave basis set is often chosen due to the periodic nature of the crystal. The pseudopotentials used should be tailored to describe the core and valence electrons of silicon accurately. Ultrasoft pseudopotentials designed for solids are commonly employed in silicon crystal simulations.

It's important to consider the convergence of the basis set and the accuracy of the pseudopotentials to ensure reliable results for properties like band structure, density of states, and electronic band gaps.

Silicon Crystal: Plane-Wave Basis Set, Ultrasoft Pseudopotentials

4.5.4 Validation and Considerations

Validation of the chosen combination of basis sets and pseudopotentials for silicon crystal should involve comparing calculated properties with experimental data, if available, and benchmarking against high-level theoretical results.

Considerations for convergence parameters, such as the energy cutoff for plane waves and the accuracy of pseudopotentials, play a significant role in the reliability of DFT calculations for solids.

4.5.5 Basis Set Superposition Error (BSSE)

In molecular calculations, the Basis Set Superposition Error (BSSE) can occur when using multiple basis sets. Researchers must carefully account for BSSE to obtain reliable results in calculations involving molecular interactions.

4.5.6 Hydrogen Atom Example

To illustrate the impact of basis sets, consider the hydrogen atom (H). Using different basis sets, such as minimal and extended sets, demonstrates how the choice affects the accuracy of electronic structure predictions.

4.5.7 Accuracy and Efficiency Trade-Off

Achieving a balance between accuracy and computational efficiency is essential. Highly accurate basis sets and pseudopotentials may be computationally expensive, and researchers must optimize their choices based on the desired level of precision.

4.5.8 Applications in Material Science

Basis sets and pseudopotentials find extensive applications in materials science, enabling the study of various materials' electronic and structural properties.

4.6 Software Packages for DFT

Several powerful software packages are available for performing Density Functional Theory (DFT) calculations. These packages provide a versatile platform for researchers and scientists to explore the electronic structure and properties of materials. In this section, we discuss some prominent DFT software packages and their features.

4.6.1   VASP (Vienna Ab initio Simulation Package)

VASP is a widely used DFT package for electronic structure calculations. It employs plane-wave basis sets and pseudopotentials, making it suitable for a range of materials. VASP is known for its accuracy in predicting materials' properties and is particularly useful for studying surfaces, defects, and electronic transitions.

4.6.2   Quantum ESPRESSO

Quantum ESPRESSO is an open-source DFT package designed for high-performance computing. It includes a suite of codes for electronic structure calculations, molecular dynamics, and quantum transport simulations. Quantum ESPRESSO is known for its efficiency and scalability on parallel architectures.

4.6.3   Gaussian

Gaussian is a comprehensive quantum chemistry software suite that includes DFT functionality. It is widely used for studying molecular systems and electronic structure calculations. Gaussian offers a user-friendly interface and supports a variety of basis sets and functionals.

4.6.4   ABINIT

ABINIT is an open-source DFT software package that employs a plane-wave basis set and pseudopotentials. It is suitable for studying a diverse range of materials, including solids, surfaces, and molecules. ABINIT focuses on efficiency and accuracy in electronic structure calculations.

4.6.5   WIEN2k

WIEN2k is a DFT package specifically designed for studying the electronic structure of solids. It uses augmented plane waves as basis sets and includes features such as spin-orbit coupling and hybrid functionals. WIEN2k is known for its accuracy in predicting band structures and total energies.

4.6.6 Example: Electronic Structure Calculation with VASP

Let's consider an example of performing an electronic structure calculation using VASP for a simple material, such as silicon (Si). We can explore the band structure and density of states to gain insights into the material's electronic properties.

4.6.7 Numerical Example: Quantum ESPRESSO for a Molecule

Suppose we want to study the electronic structure of a molecule, for instance, water (H_2O), using Quantum ESPRESSO. We can set up input files, perform a self-consistent calculation, and analyze the results, including the molecular orbitals and electronic density.

4.6.8 Comparison of Software Packages

Researchers often choose DFT software packages based on their specific research needs, computational resources, and the type of material under investigation. A comparison of the features, accuracy, and user interfaces of different packages helps in selecting the most suitable tool for a particular study.

4.6.9 Integration with High-Performance Computing (HPC)

Many DFT software packages are designed to take advantage of parallel computing capabilities, allowing researchers to perform large-scale simulations efficiently. Integration with High-Performance Computing (HPC) resources enhances the speed and scalability of calculations.

Chapter 5

Applications of DFT

5.1 Structural Optimization

Structural optimization is a fundamental application of Density Functional Theory (DFT) that enables the determination of the most stable geometric arrangement of atoms in a molecule or solid. In this section, we explore the theoretical background and practical aspects of structural optimization using DFT.

5.1.1 Theoretical Background

The goal of structural optimization is to find the equilibrium geometry of a system, where the forces acting on each atom are minimized, indicating a stable configuration. This involves iteratively adjusting atomic positions until the system reaches a minimum in the potential energy surface.

5.1.2 Basic Steps in Structural Optimization

1. **Initial Configuration:** Start with an initial atomic configuration.

2. **Energy Calculation:** Use DFT to calculate the total energy of the system based on the current atomic positions.

3. **Force Calculation:** Derive the forces acting on each atom from the energy gradient.

4. **Atomic Displacement:** Move each atom along the direction of the force to reduce potential energy.

5. **Convergence Check:** Repeat steps 2-4 until the forces and atomic positions converge to a minimum.

5.1.3   Example: Structural Optimization of a Molecule

Consider the molecule water (H_2O). To optimize its geometry, we can perform DFT calculations using a chosen functional and basis set. The iterative optimization process adjusts bond lengths and angles until the system reaches a stable configuration.

5.1.4   Numerical Example: Optimizing a Crystal Structure

For a more complex example, let's explore the structural optimization of a crystal structure, such as silicon (Si). We can investigate how changes in lattice parameters impact the material's stability and properties.

5.1.5   Challenges and Considerations

Structural optimization using DFT is a powerful tool, but it comes with challenges. Convergence criteria, choice of exchange-correlation functional, and the treatment of van der Waals interactions are factors that need careful consideration.

5.1.6   Applications in Material Science

Structural optimization plays a crucial role in material science, where understanding the stable configurations of materials is essential for predicting properties. Applications include the study of polymers, catalysts, and materials for electronic devices.

5.1.7 Integration with Experiment

DFT-optimized structures can be compared with experimental results, facilitating the validation of theoretical predictions. This integration enhances the predictive power of DFT in guiding experimental investigations.

5.2 Electronic Properties

Understanding the electronic properties of materials is a key aspect of applying Density Functional Theory (DFT). This section explores the electronic structure, band properties, and related phenomena that can be studied using DFT.

5.2.1 Electronic Structure Calculation

DFT allows for the accurate calculation of electronic structure, providing insights into the distribution of electrons in a material. The Kohn-Sham equations, solved within DFT, yield electronic density, energy levels, and wavefunctions.

5.2.2 Band Structure and Band Gap

The band structure of a material, derived from DFT calculations, reveals the arrangement of electronic energy bands. The band gap, the energy difference between the highest occupied and lowest unoccupied states, determines a material's electrical conductivity and optical properties.

5.2.3 Fermi Surface

DFT enables the visualization of the Fermi surface, representing the energy-momentum relationship of electrons at the Fermi level. This information is crucial for understanding a material's electrical and thermal transport properties.

5.2.4   Numerical Example: Band Structure of Silicon

Consider the band structure of silicon (Si). Using DFT, one can calculate the electronic band structure to identify the band gap and gain insights into the semiconductor properties of silicon.

5.2.5   Density of States (DOS)

The density of states, derived from DFT, provides a detailed account of the distribution of electronic states in a material. DOS is fundamental for understanding electronic transitions, optical properties, and thermal behavior.

5.2.6   Magnetic Properties

DFT is widely used to study the magnetic properties of materials. The calculation of magnetic moments and magnetic exchange interactions contributes to the understanding of ferromagnetism, antiferromagnetism, and other magnetic phenomena.

5.2.7   Example: Magnetic Ordering in Transition Metals

Explore the magnetic ordering of transition metals (Fe, Co, Ni) using DFT calculations. Investigate how magnetic structures influence the properties of these materials.

5.2.8   Dielectric Properties

DFT can be employed to study dielectric properties such as permittivity and susceptibility. This is valuable for understanding a material's response to external electric fields and its applications in electronic devices.

5.2.9 Example: Dielectric Constant of Insulating Materials

Calculate the dielectric constant of insulating materials like SiO_2 using DFT. Explore the relationship between electronic structure and dielectric properties.

5.2.10 Applications in Nanotechnology

DFT plays a crucial role in designing nanomaterials with tailored electronic properties. Applications include nanoelectronics, sensors, and quantum dots, where precise control over electronic structure is essential.

5.2.11 Integration with Experiment

Comparing DFT results with experimental data enhances the understanding of electronic properties. Experimental techniques such as photoemission spectroscopy validate DFT predictions and guide further investigations.

5.3 Chemical Reactions and Dynamics

Chemical reactions and dynamics play a crucial role in various scientific and industrial applications. Density Functional Theory (DFT) provides a powerful tool for investigating the energetics, mechanisms, and dynamics of chemical reactions at the atomic and molecular levels.

5.3.1 Energetics of Chemical Reactions

DFT allows the calculation of reaction energies, activation barriers, and reaction pathways. Understanding the energetics of chemical reactions is essential for predicting reaction feasibility and designing efficient catalytic processes.

5.3.2   Example: Hydrogenation of Ethylene

Consider the hydrogenation of ethylene (C_2H_4) to ethane (C_2H_6). Using DFT, calculate the reaction energy and activation barrier to elucidate the thermodynamics and kinetics of the hydrogenation process.

5.3.3   Reaction Mechanisms

DFT aids in unraveling intricate reaction mechanisms by providing insights into transition states, intermediates, and products. Visualization of reaction pathways enhances the understanding of complex chemical transformations.

5.3.4   Example: Dehydrogenation of Methane

Explore the dehydrogenation of methane (CH_4) to produce ethylene (C_2H_4). DFT calculations reveal the key intermediates and transition states involved in the reaction mechanism.

5.3.5   Surface Reactions and Catalysis

DFT is widely applied to study surface reactions on catalysts. Investigating the interaction between reactants and catalyst surfaces facilitates the design of efficient catalysts for industrial processes.

5.3.6   Example: Fischer-Tropsch Synthesis

Examine the Fischer-Tropsch synthesis, a vital process for converting syngas into hydrocarbons. DFT provides insights into the adsorption and activation of reactants on catalytic surfaces.

5.3.7   Nonadiabatic Processes

DFT-based molecular dynamics simulations enable the study of nonadiabatic processes, where electronic and nuclear motions are coupled. This is crucial for understanding ultrafast chemical reactions and excited-state dynamics.

5.3.8 Example: Photochemical Reactions

Investigate photochemical reactions using DFT to explore electronic excited states and photoinduced processes. Understanding the dynamics of photochemical reactions is essential in fields such as photophysics and photobiology.

5.3.9 Quantum Mechanical Tunneling

DFT accounts for quantum mechanical tunneling in chemical reactions, especially in hydrogen transfer reactions. Tunneling significantly influences reaction rates and is critical for accurate kinetic predictions.

5.3.10 Example: Enzymatic Reactions

Apply DFT to study enzymatic reactions where tunneling plays a crucial role. Understand how quantum effects impact the catalytic efficiency of enzymes in biological systems.

5.3.11 Integration with Experimental Techniques

Combining DFT predictions with experimental data, such as reaction kinetics and spectroscopic measurements, enhances the accuracy of reaction models. This integration is essential for validating theoretical results.

Chapter 6

Advanced Topics in DFT

6.1 Time-Dependent DFT (TDDFT)

Time-Dependent Density Functional Theory (TDDFT) extends the capabilities of standard DFT to address electronic excitations and optical properties. TDDFT is a powerful approach for studying the electronic structure of excited states in molecules and solids.

6.1.1 Basic Concepts

In TDDFT, the time evolution of the electronic density is considered to describe electronic excitations. The theory provides a framework for calculating excitation energies, oscillator strengths, and response properties.

6.1.2 Linear Response and Excitation Energies

TDDFT is based on linear response theory, where the response of the system to an external perturbation is analyzed. The excitation energies correspond to the resonances in the linear response spectrum, providing information about electronic transitions.

6.1.3   Example: UV-Visible Absorption Spectra

Calculate the UV-Visible absorption spectra of a molecule using TDDFT. Understand how the excitation energies obtained from TDDFT simulations correlate with experimental absorption peaks.

6.1.4   Optical Properties and Oscillator Strengths

TDDFT allows the prediction of optical properties, such as absorption and circular dichroism. The oscillator strengths provide insights into the intensity of electronic transitions and the nature of excited states.

6.1.5   Example: Circular Dichroism in Chiral Molecules

Explore the prediction of circular dichroism spectra for chiral molecules using TDDFT. Understand the chiroptical properties arising from the interactions of circularly polarized light with molecular systems.

6.1.6   Time-Dependent Kohn-Sham Equations

The time-dependent Kohn-Sham equations form the basis of TDDFT, where the time-dependent external potential is introduced to describe the time evolution of the electronic system. The equations are solved numerically to obtain excitation energies.

6.1.7   Example: Photoinduced Electron Transfer

Investigate photoinduced electron transfer processes using TDDFT. Study how changes in the electronic density contribute to the dynamics of charge transfer reactions.

6.1.8   Beyond Linear Response

TDDFT can be extended beyond linear response for studying non-linear optical properties and phenomena involving strong electronic excitations. Advanced

TDDFT methods address challenges associated with highly excited states.

6.1.9 Example: Two-Photon Absorption

Apply advanced TDDFT techniques to calculate two-photon absorption cross-sections. Explore the nonlinear response of materials under intense laser fields.

6.1.10 Integration with Experimental Techniques

Validate TDDFT predictions by comparing calculated spectra with experimental results. TDDFT complements experimental studies, providing a theoretical framework to interpret complex electronic and optical phenomena.

6.2 Relativistic Effects in DFT

Relativistic effects play a crucial role in accurately describing the electronic structure of heavy elements. In this section, we explore the incorporation of relativistic corrections within Density Functional Theory (DFT) and examine their impact on the calculated properties of molecules and materials.

6.2.1 Introduction to Relativistic Effects

Relativistic effects become prominent when dealing with elements with high atomic numbers. The strong magnetic field around heavy nuclei leads to significant relativistic corrections, influencing the electronic orbitals' behavior and energy levels.

6.2.2 Scalar and Spin-Orbit Corrections

Two main types of relativistic corrections are considered in DFT: scalar relativistic corrections and spin-orbit coupling. Scalar relativistic corrections account for the mass-velocity and mass-contraction effects, while spin-orbit coupling involves the interaction between the electronic spin and the orbital motion.

6.2.3   Example: Heavy Metal Complexes

Investigate the impact of relativistic effects on the electronic structure of heavy metal complexes, such as platinum or gold-containing compounds. Compare DFT calculations with and without relativistic corrections to understand the differences in bond lengths and energies.

6.2.4   Fully Relativistic Approaches

Fully relativistic DFT methods, such as the four-component Dirac-Kohn-Sham equations, provide a comprehensive treatment of relativistic effects. These methods explicitly consider the electron's spin and angular momentum, offering accurate descriptions of heavy element systems.

6.2.5   Example: Prediction of Electronic Spectra

Apply fully relativistic DFT methods to predict electronic spectra of lanthanide complexes. Explore how spin-orbit coupling influences the electronic transitions and absorption/emission energies in these systems.

6.2.6   Effect on Bonding and Reactivity

Relativistic corrections influence the bonding characteristics and reactivity of heavy elements. Understand how relativistic effects modify bond strengths, bond angles, and reaction pathways in compounds containing heavy atoms.

6.2.7   Example: Bonding in Actinide Compounds

Examine the bonding in actinide compounds and uranium complexes using relativistic DFT. Investigate how scalar and spin-orbit corrections affect the stability and structure of these compounds.

6.2.8 Implementation in Computational Codes

Discuss the implementation of relativistic corrections in popular DFT software packages. Highlight considerations for choosing appropriate computational settings to account for relativistic effects in simulations.

6.2.9 Comparison with Experimental Data

Validate relativistic DFT results by comparing calculated properties with experimental data for heavy element systems. Assess the accuracy of the chosen computational approach in reproducing experimental observables.

6.2.10 Applications in Materials Science

Explore the applications of relativistic DFT in materials science, including the design of novel materials with specific electronic and magnetic properties. Understand how relativistic effects impact the stability and performance of materials.

6.3 DFT for Strongly Correlated Systems

Density Functional Theory (DFT) has proven to be a powerful tool for studying electronic structures, but it faces challenges in accurately describing strongly correlated systems. In this section, we delve into the complexities of such systems and explore strategies within DFT to handle strong electron-electron interactions.

6.3.1 Introduction to Strongly Correlated Systems

Strongly correlated systems are characterized by a significant interplay between electron-electron interactions and quantum fluctuations. Traditional DFT methods may struggle to capture the complex electronic behaviors observed in these systems, making specialized approaches necessary.

6.3.2   Hubbard U Correction

One common strategy to address electron correlation effects is the Hubbard U correction. This correction introduces an on-site Coulomb repulsion term, U, to account for the localized nature of electron-electron interactions. Explore the impact of Hubbard U on the electronic structure of transition metal compounds.

6.3.3   Example: Transition Metal Oxides

Investigate the electronic structure of transition metal oxides, such as VO_2 or Fe_2O_3, using DFT with and without Hubbard U correction. Compare the results to experimental data and understand how the inclusion of U affects the predicted properties.

6.3.4   DFT+U Methodology

Discuss the DFT+U methodology in detail, focusing on the selection of the Hubbard U parameter and its impact on different electronic states. Provide guidelines for choosing appropriate U values for various materials and discuss the limitations of this approach.

6.3.5   Quantum Monte Carlo (QMC) Methods

Introduce Quantum Monte Carlo (QMC) methods as an alternative approach to study strongly correlated systems. QMC techniques, such as the Diffusion Monte Carlo method, provide a more accurate treatment of electron correlation but come with their own computational challenges.

6.3.6   Example: Correlated Electron Liquids

Apply Quantum Monte Carlo methods to study correlated electron liquids and Mott insulators. Investigate the emergence of charge-density waves, spin fluctuations, and other correlated phenomena in these systems.

6.3.7 Cluster Dynamical Mean-Field Theory (CDMFT)

Explore advanced techniques like Cluster Dynamical Mean-Field Theory (CDMFT) to go beyond local approximations and capture non-local correlation effects. Understand the role of cluster size in improving the description of correlation effects.

6.3.8 Example: CDMFT in Molecular Systems

Apply Cluster Dynamical Mean-Field Theory to molecular systems with strong electron correlation. Analyze how non-local correlation effects influence the electronic and magnetic properties of molecular complexes.

6.3.9 Machine Learning Approaches

Discuss the emerging role of machine learning approaches in addressing electron correlation challenges. Explore how neural networks and other machine learning techniques can assist in predicting electronic properties of strongly correlated materials.

6.3.10 Example: Predicting Correlation Effects

Demonstrate the application of machine learning to predict correlation effects in materials. Showcase how these methods can accelerate the exploration of new materials with desired electronic properties.

6.3.11 Challenges and Future Directions

Highlight the challenges and ongoing research in the field of DFT for strongly correlated systems. Discuss potential future directions, including the development of hybrid methods and the integration of machine learning into DFT calculations.

Chapter 7

DFT in Materials Science

7.1 Electronic Structure of Solids

Understanding the electronic structure of solids is crucial for predicting material properties and designing novel materials. Density Functional Theory (DFT) provides a powerful framework for studying the electronic properties of solids.

7.1.1 Theoretical Foundations

The electronic structure of solids is governed by the distribution of electrons in the crystal lattice. DFT allows for the calculation of electronic band structures, density of states, and other key electronic properties.

7.1.2 Example: Band Structure of Silicon

Consider the example of silicon, a semiconductor widely used in electronic devices. Use DFT to calculate the band structure of silicon and analyze the energy bands, band gap, and effective mass of charge carriers.

DFT Calculations for Silicon Band Structure

Perform DFT calculations to obtain the electronic band structure of silicon. Visualize the energy bands and analyze the band gap, which is crucial for understanding the semiconductor behavior of silicon.

7.1.3   Numerical Example: Density of States

The density of states (DOS) is a key quantity in electronic structure calculations. DFT allows for the computation of the DOS, providing insights into the available electronic states at different energy levels.

DFT Calculations for Density of States

Calculate the density of states for a specific material using DFT. Analyze how the DOS varies with energy and understand the relationship between electronic states and material properties.

7.1.4   Chemical Bonding in Solids

DFT can elucidate the nature of chemical bonding in solids, ranging from metallic bonding in metals to covalent bonding in semiconductors and insulators.

Example: Covalent Bonding in Diamond

Explore the covalent bonding in diamond using DFT calculations. Investigate the electronic structure of carbon-carbon bonds and understand how it contributes to diamond's unique properties.

7.1.5   Advances in Electronic Structure Calculations

Recent advancements in DFT methodologies have led to more accurate predictions of electronic structure properties, including the incorporation of van der Waals interactions and the treatment of strongly correlated electron systems.

Example: Van der Waals Interactions in Layered Materials

Apply advanced DFT methods to study the role of van der Waals interactions in layered materials. Examine how these interactions influence the electronic structure and stability of materials like graphene.

7.2 DFT Applications in Nanomaterials

The application of Density Functional Theory (DFT) to nanomaterials has revolutionized our understanding and manipulation of materials at the nanoscale. This section explores various applications, providing sample examples and numerical analyses.

7.2.1 Theoretical Framework for Nanomaterials

Nanomaterials exhibit unique properties due to their small size and high surface-to-volume ratio. DFT offers a theoretical framework to investigate electronic, optical, and structural properties of nanomaterials.

Mathematical Formulation for Nanoscale Effects

Consider the mathematical formulation of quantum confinement effects in nanomaterials. Use DFT to model the size-dependent behavior of electronic states, band gaps, and optical properties in nanoparticles.

7.2.2 Example: Quantum Dots for Optoelectronic Devices

Explore the application of DFT in designing quantum dots for optoelectronic devices. Perform DFT calculations to predict electronic band structures, energy levels, and absorption spectra of quantum dots.

Numerical Analysis: Tuning Band Gaps in Quantum Dots

Perform numerical analyses to understand how the size and composition of quantum dots influence their band gaps. Investigate strategies for tuning band gaps to optimize quantum dot performance in devices.

7.2.3   Chemical Functionalization of Nanomaterials

DFT enables the study of chemical functionalization on nanomaterial surfaces. Investigate the adsorption of molecules on nanotubes or nanoparticles, considering electronic and structural changes.

Example: Functionalization of Carbon Nanotubes

Study the functionalization of carbon nanotubes with various chemical groups. Use DFT to analyze the impact on electronic structure, mechanical properties, and potential applications in nanodevices.

7.2.4   Quantum Transport in Nanostructures

DFT plays a crucial role in understanding quantum transport phenomena in nanoscale systems. Investigate electronic transport properties of nanowires, nanotubes, and graphene-based devices.

Example: Conductance in Molecular Junctions

Apply DFT to calculate the conductance of molecular junctions. Explore how different molecular structures affect electron transport and study the feasibility of molecular-scale electronics.

7.2.5   Advanced Techniques for Nanomaterial Simulations

Recent advancements in DFT methodologies allow for more accurate simulations of nanomaterials. Explore the use of hybrid functionals, time-dependent DFT, and machine learning techniques in nanomaterial studies.

Example: Machine Learning for Nanomaterial Design

Integrate machine learning approaches with DFT to accelerate nanomaterial discovery. Train models to predict material properties, allowing for efficient exploration of vast nanomaterial design spaces.

In summary, DFT applications in nanomaterials span a wide range of topics, from quantum confinement effects to functionalization and quantum transport. By providing examples and numerical analyses, this section aims to illustrate the versatility and power of DFT in nanomaterial research.

7.3 DFT in Surface Science

Surface science is a critical domain for applying Density Functional Theory (DFT) to unravel the properties of materials at interfaces. This section delves into theoretical foundations, adsorption phenomena, surface reactivity, and mathematical formulations inherent in DFT calculations for surface science.

7.3.1 Theoretical Foundations

In surface science applications of DFT, the theoretical foundation involves modeling surface energy, electronic structure, and reactivity. Surface energy calculations using DFT play a crucial role in understanding the stability and growth of crystal facets.

Surface Energy Calculations

To compute surface energies, we employ DFT to analyze various crystallographic facets. The surface energy influences the stability of crystals and has profound implications for material properties.

7.3.2 Adsorption on Surfaces

DFT is a powerful tool for studying the adsorption of molecules on surfaces. This includes predicting adsorption energies, understanding binding geometries, and exploring applications in sensing and catalysis.

Example: Adsorption of Gases on Metal Surfaces

Consider the adsorption of gases, such as hydrogen (H_2) or oxygen (O_2), on metal surfaces. Employ DFT calculations to predict adsorption energies, binding configurations, and the influence on surface reactivity.

Numerical Analysis: Coverage Dependence of Adsorption

Perform numerical analyses to investigate how the coverage of adsorbates impacts their interactions on surfaces. Explore saturation coverage and study the competitive adsorption of multiple species.

7.3.3 Surface Reactivity and Catalysis

DFT provides insights into surface reactivity, particularly in catalytic processes. Explore catalytic mechanisms, identify active sites, and optimize catalytic materials.

Example: Oxygen Reduction Reaction on Metal Surfaces

Investigate the oxygen reduction reaction (ORR) on metal surfaces, a critical process in fuel cell applications. Utilize DFT to elucidate reaction pathways, predict intermediates, and assess catalytic activity.

7.3.4 Mathematical Formulations

DFT calculations for surface science involve mathematical formulations, considering periodic boundary conditions, slab models, and the treatment of van der Waals interactions in heterogeneous systems.

Equations: Surface Energy and Adsorption

Incorporate relevant mathematical equations, such as expressions for surface energy and adsorption energy equations. Highlight the role of electronic structure and density in these formulations.

7.3.5 Visualization: Bonding Diagrams

Use chemfig to create clear and illustrative bonding diagrams that depict interactions between adsorbates and surfaces.

Example: Bonding in Adsorbed Molecules

Create a bonding diagram illustrating interactions between an adsorbate and a metal surface. Emphasize changes in electronic structure and bond formation during adsorption.

Chapter 8

Environmental Applications of DFT

8.1 DFT in Catalysis

Density Functional Theory (DFT) has emerged as a powerful tool in understanding and optimizing catalytic processes. The application of DFT to catalysis provides valuable insights into reaction mechanisms, selectivity, and the design of efficient catalysts.

8.1.1 Theoretical Foundations

In DFT, the total energy of a system is expressed as a functional of the electron density. The fundamental equation governing DFT is the Kohn-Sham equation:

$$\left[-\frac{\hbar^2}{2m} \nabla^2 + V_{\text{eff}}(\mathbf{r}) \right] \psi_i(\mathbf{r}) = \varepsilon_i \psi_i(\mathbf{r}) \tag{8.1}$$

where $\psi_i(\mathbf{r})$ is the Kohn-Sham orbital, $V_{\text{eff}}(\mathbf{r})$ is the effective potential, and ε_i is the orbital energy.

8.1.2   Catalytic Reaction Example: Hydrogenation

Consider the catalytic hydrogenation of ethylene (C_2H_4) on a metal surface. The reaction involves the adsorption of H_2 and subsequent bond formation:

$$C_2H_4 + H_2 \longrightarrow C_2H_6 \tag{8.2}$$

DFT can be employed to calculate the energy profiles of different reaction pathways, identify transition states, and predict reaction kinetics.

DFT Calculations for Hydrogenation

Perform DFT calculations to determine the adsorption energies of C_2H_4, H_2, and the reaction intermediate $C_2H_4H_2$ on the metal surface. The Gibbs free energy change (ΔG) can be calculated to assess the thermodynamic feasibility.

8.1.3   Numerical Example: Palladium-Catalyzed Suzuki-Miyaura Cross-Coupling

Consider the Suzuki-Miyaura cross-coupling reaction catalyzed by palladium (Pd). The reaction involves the coupling of an aryl halide ($Ar-X$) with an organoboron compound ($B-R$):

$$Ar-X + B-R \longrightarrow Ar-B + H-X \tag{8.3}$$

DFT can elucidate the reaction mechanism, optimize reaction conditions, and predict the selectivity of the coupling partners.

DFT Calculations for Suzuki-Miyaura Cross-Coupling

Perform DFT calculations to investigate the oxidative addition, transmetalation, and reductive elimination steps of the catalytic cycle. Analyze the energy barriers and reaction intermediates to gain mechanistic insights.

8.1.4 Challenges and Future Directions

Despite its successes, DFT in catalysis faces challenges such as accurate treatment of dispersion forces and the need for efficient treatment of large catalytic systems.

Example: Addressing Dispersion Forces

Explore dispersion-corrected DFT methods to improve the accuracy of predictions for catalytic systems involving weak interactions.

8.2 Environmental Impact Assessments

Density Functional Theory (DFT) plays a crucial role in environmental impact assessments by providing a theoretical framework to analyze the interactions between pollutants and environmental systems.

8.2.1 Theoretical Foundations

The mathematical formulation of DFT allows for the calculation of electronic structure and properties of molecules and materials. In environmental applications, DFT is employed to model the behavior of pollutants, understand their reactivity, and assess their impact on ecosystems.

8.2.2 Example: Adsorption of Heavy Metals

Consider the adsorption of heavy metals such as mercury (Hg) on soil particles. DFT can be used to calculate adsorption energies, study surface interactions, and predict the mobility of heavy metal pollutants in soil.

DFT Calculations for Heavy Metal Adsorption

Perform DFT calculations to investigate the adsorption of Hg on various soil surfaces. Analyze the electronic structure and energetics of the adsorption process to assess the environmental risk.

8.2.3   Numerical Example: Water Contaminant Removal

DFT is instrumental in designing materials for water purification. The removal of contaminants like arsenic (As) from water sources can be studied using DFT.

DFT Calculations for Arsenic Removal

Perform DFT calculations to explore the interactions between arsenic-contaminated water and functionalized materials. Assess the feasibility and efficiency of different materials for arsenic removal.

8.2.4   Challenges and Future Directions

Despite its successes, applying DFT to environmental impact assessments faces challenges such as the need for accurate parameterization of environmental systems and consideration of dynamic processes.

Example: Dynamic Environmental Processes

Explore the integration of molecular dynamics (MD) simulations with DFT to model dynamic processes in environmental systems, such as pollutant transport in groundwater.

8.3   DFT in Gas Adsorption Studies

Gas adsorption studies using Density Functional Theory (DFT) provide valuable insights into the interaction between gases and porous materials. This section explores the theoretical foundations and applications of DFT in understanding and predicting gas adsorption behavior.

8.3.1   Theoretical Foundations

The mathematical formulation of DFT enables the calculation of adsorption energies, binding sites, and isotherms for gases on various surfaces. This information is crucial for designing materials with enhanced gas adsorption properties.

8.3.2 Numerical Example: Methane (CH_4) Storage

In this numerical example, we explore the application of density functional theory (DFT) to understand the storage of methane (CH_4). Methane storage is of significant interest due to its potential as a clean energy source. We will utilize DFT calculations to investigate the adsorption of methane on a porous material.

Selection of Porous Material

Choosing an appropriate porous material is crucial for efficient methane storage. We consider a metal-organic framework (MOF) known for its high surface area and tunable properties. The MOF under investigation is denoted as MOF-XYZ.

Adsorption Isotherm

We perform DFT calculations to generate the adsorption isotherm of methane on MOF-XYZ. The isotherm illustrates the relationship between the amount of adsorbed methane and the pressure at different temperatures. This information helps in determining the storage capacity of MOF-XYZ under varying conditions.

Energetics of Adsorption

DFT allows us to investigate the energetics of methane adsorption on MOF-XYZ. By calculating the adsorption energy, we gain insights into the strength of the interaction between methane molecules and the MOF. Understanding the energetics is crucial for assessing the stability of the adsorbed methane and predicting release conditions.

Comparative Analysis

To validate our results, we compare the DFT-calculated adsorption isotherm and energetics with experimental data, if available. This comparative analysis helps in assessing the accuracy of the DFT predictions and provides confidence in the reliability of the computational approach.

Impact of Temperature

Finally, we explore the impact of temperature on methane storage. DFT calculations at different temperatures allow us to analyze how storage capacity and energetics vary under realistic operating conditions.

8.3.3   Numerical Example: Methane (CH_4) Storage

In this numerical example, we explore the application of density functional theory (DFT) to understand the storage of methane (CH_4). Methane storage is of significant interest due to its potential as a clean energy source. We will utilize DFT calculations to investigate the adsorption of methane on a porous material.

Selection of Porous Material

Choosing an appropriate porous material is crucial for efficient methane storage. We consider a metal-organic framework (MOF) known for its high surface area and tunable properties. The MOF under investigation is denoted as MOF-XYZ.

Adsorption Isotherm

We perform DFT calculations to generate the adsorption isotherm of methane on MOF-XYZ. The isotherm illustrates the relationship between the amount of adsorbed methane and the pressure at different temperatures. This information helps in determining the storage capacity of MOF-XYZ under varying conditions.

Energetics of Adsorption

DFT allows us to investigate the energetics of methane adsorption on MOF-XYZ. By calculating the adsorption energy, we gain insights into the strength of the interaction between methane molecules and the MOF. Understanding the energetics is crucial for assessing the stability of the adsorbed methane and predicting release conditions.

Comparative Analysis

To validate our results, we compare the DFT-calculated adsorption isotherm and energetics with experimental data, if available. This comparative analysis helps in assessing the accuracy of the DFT predictions and provides confidence in the reliability of the computational approach.

Impact of Temperature

Finally, we explore the impact of temperature on methane storage. DFT calculations at different temperatures allow us to analyze how storage capacity and energetics vary under realistic operating conditions.

In conclusion, this numerical example demonstrates the power of DFT in understanding and predicting methane storage on porous materials. Such studies contribute to the development of efficient and sustainable energy storage solutions.

DFT Calculations for Methane Storage

Perform DFT calculations to investigate the adsorption of methane on different materials. Explore the impact of pore size, temperature, and pressure on methane storage capacity.

8.3.4 Challenges and Advances

Gas adsorption studies using DFT face challenges such as accurately describing the dynamics of gas molecules and accounting for temperature and pressure effects. Recent advances in DFT methods address these challenges, allowing for more realistic simulations.

Example: Temperature and Pressure Effects

Incorporate temperature and pressure effects into DFT simulations to better model real-world gas adsorption processes. Explore the role of thermal fluctuations and dynamic gas interactions.

Chapter 9

Emerging Trends and Future Perspectives

9.1 Machine Learning in DFT

The integration of Machine Learning (ML) with Density Functional Theory (DFT) has revolutionized computational materials science. ML algorithms enhance the efficiency and accuracy of DFT calculations, offering a promising avenue for tackling complex problems.

9.1.1 Overview of ML-DFT Hybrid

In ML-DFT hybrid approaches, a model is trained on a dataset of DFT-calculated properties. Once trained, the model can predict material properties with reduced computational cost, enabling high-throughput screening of materials.

Mathematical Formulation of ML-DFT Model

Let $\mathbf{X}$ represent the input features (structural descriptors) and $\mathbf{Y}$ the DFT-calculated properties. The ML-DFT model can be defined as:

$$\mathbf{Y} = f(\mathbf{X}) + \epsilon \tag{9.1}$$

where $f(\cdot)$ is the ML model, and ϵ accounts for the prediction error.

9.1.2 Sample Working Example: Band Gap Prediction

Consider a ML-DFT model trained on a dataset of band gaps of various semi-conductors. The model can predict the band gap of a new material based on its structural features.

```
# Python code using scikit-learn
from sklearn.model_selection import train_test_split
from sklearn.ensemble import RandomForestRegressor


# Load DFT-calculated data
X, Y = load_dft_data()


# Split data into training and testing sets
X_train, X_test, Y_train, Y_test =
train_test_split(X, Y, test_size=0.2)


# Train the ML model
ml_model = RandomForestRegressor()
ml_model.fit(X_train, Y_train)


# Predict band gap for a new material
new_material_features = extract_features(new_material_structure)
predicted_band_gap = ml_model.predict(new_material_features)
```

9.1.3 Numerical Example: Accelerated Geometry Optimization

ML can accelerate geometry optimization by predicting the energy landscape. A ML model trained on DFT-calculated energies can guide the optimization algorithm to converge faster.

Consider a molecular structure optimization using ML-DFT:

```python
# Python code using TensorFlow and ASE
import tensorflow as tf
from ase import Atoms
from ase.optimize import BFGS

# Load DFT-calculated energy data
X, Y_energy = load_dft_energy_data()

# Train the ML model
ml_energy_model = tf.keras.models.Sequential([
    tf.keras.layers.Dense(units=64, activation='relu'),
    tf.keras.layers.Dense(units=1)
])
ml_energy_model.compile(optimizer='adam', loss=
'mean_squared_error')
ml_energy_model.fit(X, Y_energy, epochs=50)

# Perform ML-accelerated geometry optimization
initial_structure = Atoms('H2', positions=[[0, 0, 0], [0, 0, 0.7]])
ml_optimization = BFGS(initial_structure, trajectory='optimized.traj',
                       logfile='optimization.log', force_consistent=True)
ml_optimization.run(fmax=0.01)
```

9.1.4 Challenges and Future Directions

While ML-DFT offers remarkable advantages, challenges such as interpretability, data quality, and model transferability persist. The interpretability of ML models remains an active area of research.

Example: Addressing Model Interpretability

Explore techniques like SHAP (SHapley Additive exPlanations) values to enhance the interpretability of ML models. SHAP values provide insights into the
contribution of each feature to the model's predictions.

9.2   Advancements in DFT Methods

The continuous evolution of Density Functional Theory (DFT) has led to numerous advancements, pushing the boundaries of accuracy and efficiency. This
section explores key developments and their impact on electronic structure calculations.

9.2.1   Hybrid Functionals: A Blend of Accuracy and Effi-
ciency

Hybrid functionals combine Hartree-Fock exchange with DFT exchange-correlation
functionals, addressing the limitations of traditional DFT. The hybrid functional
is defined as:

$$E_{\text{hybrid}} = a \cdot E_{\text{HF}} + (1 - a) \cdot E_{\text{DFT}} \tag{9.2}$$

where E_{HF} is the Hartree-Fock energy, E_{DFT} is the DFT energy, and a is the
mixing parameter.

Example: Band Gap Prediction in Semiconductors

Consider the application of the HSE06 hybrid functional for band gap predictions in semiconductors. The HSE06 functional often provides more accurate
results compared to standard DFT functionals, especially for materials with
challenging electronic structures.

9.2.2 Meta-GGA Functionals: Beyond Gradient Corrections

Meta-GGA functionals introduce kinetic energy density as an additional variable, offering improvements in describing non-local electron density effects. The TPSS functional is an example, defined as:

$$E_{\text{TPSS}} = E_{\text{GGA}} + E_{\text{non-local}} \tag{9.3}$$

where E_{GGA} is the Generalized Gradient Approximation energy and $E_{\text{non-local}}$ captures non-local contributions.

Example: Molecular Properties with Meta-GGA

Explore the application of meta-GGA functionals in predicting molecular properties. The TPSS functional is known for its accuracy in describing molecular geometries and reaction pathways.

9.2.3 Time-Dependent DFT (TDDFT): Unraveling Excited States

TDDFT extends DFT to study electronic excitations and excited states. The excitation energy is determined by solving the TDDFT equations:

$$\hat{H}_{\text{TDDFT}}\Psi_i = \omega_i \Psi_i \tag{9.4}$$

where $\hat{H}_{\text{TDDFT}}$ is the TDDFT Hamiltonian, Ψ_i is the wavefunction, and ω_i is the excitation energy.

Example: UV-Vis Spectra Prediction

Apply TDDFT to predict UV-Vis absorption spectra. TDDFT calculations provide insights into electronic transitions, aiding in the interpretation of experimental spectra for various compounds.

9.2.4   Dispersion-Corrected DFT: Van der Waals Interactions

Traditional DFT struggles with van der Waals interactions, leading to inaccuracies in molecular and material properties. Dispersion-corrected DFT methods, such as DFT-D3, include empirical corrections to capture van der Waals forces.

Example: Adsorption on Surfaces

Illustrate the significance of dispersion corrections in adsorption studies. Compare results with and without dispersion corrections to emphasize the impact on surface interactions.

9.2.5   Machine Learning Accelerated DFT: Bridging Speed and Accuracy

Machine Learning techniques, including neural networks, are integrated with DFT to accelerate calculations. The ML model is trained on a dataset of DFT results to predict properties with reduced computational cost.

Example: Accelerated Geometry Optimization

Demonstrate the use of Machine Learning to expedite geometry optimization. By training the model on a diverse set of molecular structures, the ML-accelerated DFT approach speeds up the convergence of geometry optimization.

9.2.6   Challenges and Future Directions

Despite these advancements, challenges persist. Strongly correlated systems, accurate treatment of excited states, and the development of non-empirical functionals are areas that demand ongoing research. The future of DFT involves exploring new mathematical frameworks and harnessing the potential of quantum computers.

Example: Challenges in Modeling Correlated Electron Systems

Examine the challenges associated with modeling strongly correlated electron systems using DFT. Current limitations and potential avenues for overcoming these challenges should be explored through collaborative research efforts.

9.3 Interdisciplinary Applications

The application of Density Functional Theory (DFT) extends beyond traditional materials science, finding interdisciplinary relevance in various scientific domains. This section explores the diverse applications of DFT across disciplines, providing examples that showcase its versatility.

9.3.1 Catalysis in Environmental Chemistry

One notable interdisciplinary application of DFT is in environmental chemistry, particularly in catalysis. DFT calculations enable the study of catalytic mechanisms and the design of more efficient catalysts for environmental remediation processes.

Chemical Equation: Catalytic Decomposition

Utilize for a chemical equation illustrating the catalytic decomposition of a pollutant. Showcase how DFT insights guide the selection of catalysts and predict reaction pathways to minimize environmental impact.

9.3.2 Biophysical Applications in Biology

DFT plays a crucial role in understanding molecular interactions in biological systems. Explore its applications in predicting protein-ligand binding energies, elucidating reaction mechanisms in enzymatic processes, and contributing to drug discovery.

Mathematical Formulation: Protein-Ligand Binding Energy

Present a mathematical formulation for predicting protein-ligand binding energy using DFT-derived parameters. Highlight the significance of accurate energy predictions in drug design and development.

9.3.3   Energy Storage Materials in Physics

In the realm of physics, DFT is instrumental in studying materials for energy storage applications. Investigate its use in predicting the electronic structure and properties of materials for batteries and capacitors.

Bonding Diagram: Lithium-Ion Battery

Create a bonding diagram using to visualize the electronic structure of a lithium-ion battery material. Illustrate the role of DFT in optimizing the arrangement of atoms for enhanced energy storage.

9.3.4   Mathematical Modeling in Engineering

DFT finds applications in engineering disciplines by providing insights into material properties. Showcase its role in mathematical modeling of structural materials, helping engineers optimize designs for strength and durability.

Numerical Example: Stress-Strain Analysis

Provide a numerical example demonstrating stress-strain analysis using DFT-derived material properties. Illustrate how such analyses contribute to the development of robust engineering materials.

Appendix

9.4 Useful Constants

This appendix provides a compilation of useful constants frequently encountered in Density Functional Theory (DFT) calculations. These constants play a crucial role in various equations and formulations used in the DFT framework.

1. **Planck's Constant** (h): Planck's constant represents the quantum of action in quantum mechanics and is fundamental in DFT calculations.

$$h = 6.626 \times 10^{-34} \, \text{J} \cdot \text{s}$$

2. **Speed of Light** (c): The speed of light is a fundamental constant in electromagnetic theory and relativistic DFT calculations.

$$c = 2.998 \times 10^{8} \, \text{m/s}$$

3. **Avogadro's Number** (N_A): Avogadro's number is the number of atoms, ions, or molecules in one mole of a substance.

$$N_\text{A} = 6.022 \times 10^{23} \, \text{mol}^{-1}$$

4. **Boltzmann Constant** (k_B): The Boltzmann constant relates the average kinetic energy of particles in a gas to the temperature of the gas.

$$k_\text{B} = 1.380 \times 10^{-23} \, \text{J/K}$$

5. **Electron Mass (m_e):** The electron mass is a fundamental constant used in electronic structure calculations.

$$m_e = 9.109 \times 10^{-31} \, \text{kg}$$

6. **Elementary Charge (e):** The elementary charge represents the electric charge of a single electron.

$$e = 1.602 \times 10^{-19} \, \text{C}$$

7. **Permittivity of Free Space (ε_0):** The permittivity of free space is a constant that characterizes the ability of a vacuum to permit electric field lines.

$$\varepsilon_0 = 8.854 \times 10^{-12} \, \text{F/m}$$

These constants are essential in formulating and solving various equations in DFT, providing a foundation for accurate and meaningful calculations. They ensure consistency and accuracy in the application of DFT principles across different systems and scenarios.

9.5   Computational Tools and Resources

This appendix provides an overview of essential computational tools and resources commonly utilized in Density Functional Theory (DFT) calculations. These tools play a crucial role in performing accurate and efficient DFT simulations. Below are some key components:

9.5.1   Quantum Espresso

Quantum Espresso is an integrated suite of computer codes for electronic structure calculations and materials modeling. It is widely used in DFT simulations to predict the properties of materials at the atomic scale. Here is a basic input script for a Quantum Espresso calculation:

```
&CONTROL
  calculation = 'scf'
  prefix = 'example'
  pseudo_dir = '/path/to/pseudopotentials'
/
&SYSTEM
  ibrav = 2
  celldm(1) = 10.0
  nat = 2
  ntyp = 1
  ecutwfc = 30.0
  occupations = 'smearing'
  smearing = 'mp'
  degauss = 0.02
/
&ELECTRONS
  mixing_beta = 0.7
/
ATOMIC_SPECIES
  element_name  atomic_mass  pseudopotential_file
  X             28.09        X.pbe-mt_fhi.UPF
ATOMIC_POSITIONS
  X 0.0 0.0 0.0
  X 0.25 0.25 0.25
K_POINTS
  8
  0.5 0.5 0.5 1.0
```

This script performs a self-consistent field (SCF) calculation for a simple crystal.

9.5.2   VASP

The Vienna Ab initio Simulation Package (VASP) is another powerful software package for electronic structure calculations. It employs plane-wave basis sets and pseudopotentials to achieve accurate results. A typical VASP input file looks like:

```
SYSTEM = Example
ISMEAR = 0
SIGMA = 0.1
ENCUT = 400
PREC = Accurate
IALGO = 38
```

This input sets the system name, smearing method, energy cutoff, precision, and algorithm.

9.5.3   Materials Project Database

The Materials Project is a valuable online resource that provides access to a vast database of calculated materials properties. Researchers can explore the database to access computed properties for various materials, helping in the selection of materials for specific applications.

These computational tools and resources are integral to the practice of DFT and contribute significantly to the advancement of materials science and related fields.

Books

1. Kohn, W., & Sham, L. J. (1965). Self-consistent equations including exchange and correlation effects. *Physical Review*, 140(4A), A1133.

2. Martin, R. M. (2004). *Electronic Structure: Basic Theory and Practical Methods*. Cambridge University Press.

3. Parr, R. G., & Yang, W. (1989). *Density-Functional Theory of Atoms and Molecules.* Oxford University Press.

Journal Articles

1. Hohenberg, P., & Kohn, W. (1964). Inhomogeneous Electron Gas. *Physical Review*, 136(3B), B864.

2. Perdew, J. P., Burke, K., & Ernzerhof, M. (1996). Generalized Gradient Approximation Made Simple. *Physical Review Letters*, 77(18), 3865.

3. Ceperley, D. M., & Alder, B. J. (1980). Ground State of the Electron Gas by a Stochastic Method. *Physical Review Letters*, 45(7), 566.

Online Resources

1. Quantum Espresso. (n.d.). Retrieved from `https://www.quantum-espresso.org/`

2. Materials Project. (n.d.). Retrieved from `https://materialsproject.org/`

3. VASP - Vienna Ab initio Simulation Package. (n.d.). Retrieved from `https://www.vasp.at/`

Conference Proceedings

1. Kresse, G., & Furthmüller, J. (1996). Efficiency of Ab-Initio Total Energy Calculations for Metals and Semiconductors Using a Plane-Wave Basis Set. *Computational Materials Science*, 6(1), 15-50.

2. Giannozzi, P., Baroni, S., Bonini, N., Calandra, M., Car, R., Cavazzoni, C., ... & Kokalj, A. (2009). QUANTUM ESPRESSO: a modular and open-source software project for quantum simulations of materials. *Journal of Physics: Condensed Matter*, 21(39), 395502.